# STATISTICS

# FOR

# BEGINNERS

How easy it is to lie with statistics in business,

social science, politics, criminology and law

# DISCLAIMER

# Table of Contents

# Introduction

You wake up one morning to Alexa playing the music you love. You groggily step forward towards the bathroom where the mirror shows you the weather outside. As you get ready, the breakfast downstairs has just finished cooking. Your coffee has already been brewed and poured for you. All you need to do is go downstairs and enjoy a nice breakfast before you get in your car, that drives you to work by itself.

This is what's known as The Internet of Things. The proposed future of what the internet will be. A world where you have the ease and luxury of many sci-fi films. Do you know what powers such technology? The world of statistics and machine learning. It is statistics in which the computers learn what time you get up, what type of food you like, what temperature you like, and the list goes on. It's also what monitors your health as yourself and as a society. It's what allows cars to drive themselves thanks to the probability matrices. In a world that is powered by statistics and machines that learn based on those statistics, how dangerous is a statistic that lies?

## You Cannot Escape Statistics

### Statistics is How Society Works

You can't really escape statistics because it pretty much decides how laws and rules are made. In the past, laws were primarily made based off what people liked and disliked. This was due to religion most of the time. What most people were afraid of or disliked the most is what was made illegal. Therefore, if people thought that dancing was a work of the devil then it would be outlawed.

The problem is a lot of people like dancing and some of the things that were illegal, should not have been illegal. This is where you have several different problems line up in a row. You can't make laws on subjectivity. Subjectivity is where the law is subjective to the circumstance that it was created in. Therefore, something like not singing on the beach with your shoes off in Florida is subjective. Many of the strange laws that we have from the past were based on subjectivity, which is to say the circumstance is what created them. The laws don't really make sense, and no one can logically justify it. However, there are some subjective laws that persists in the world. These are mainly in countries that are heavily ideological. While I'm not going to get into the details of these countries, I will say that the more technologically advanced societies have moved vastly away from subjective laws and are almost always relying on statistics.

## Statistics Run Our Medicine

I think the best place to exemplify where statistics are used the most is in medicine. Medicine is literally decided by statistics. Let's say that you wanted to release a new drug on the market. You wanted to find a drug that could cure nail biting. The first thing that you would do is you would hire a bunch of researchers (or you would research it yourself) but ultimately you would want to gain statistics on why people bite their nails. You couldn't even begin your project without learning the why, which is defined by statistics.

However, let's go into a little bit more detail. Once you figure out the why, you then want to create methods for determining the best case of action or medicine. You would want to run a sampling group of different medicines that were known to help with nail biting. In this sampling group, you would want to measure which one was the most effective and that would be done with statistics. I could go on and on, but I think you see the point. Medicine is defined by how successful it is and that is defined by statistics.

**Statistics Make Our Laws**

I think the best-case example of where statistics defines our laws is in the case of child safety. Most of the time you hear the gun laws being associated with the violence against children. In fact, many of these guns that are banned have been banned because of how effective they are at causing death. The only problem is that it's a bit flawed.

However, there are more notable examples that are less known. For instance, do you know why the asphalt material is used for roads? The companies and government agencies employed to pave the roads had to figure out a cheaper way to repair and maintain roads. As we've noted in the medicine section, the success of an item versus other items is defined by statistics. We used to use plain old dirt and then we moved up to rocks and then bricks only to finally land in asphalt. Even today, they are still trying to use to statistics to find better ways or rather cheaper ways to maintain roads. However, the most important bit is that they do so using statistics.

**The Importance of Statistics**

**The World of IOT**

We are going to have a lot of devices on the market. A lot of devices that will be connected to the internet. The internet of things, as it is referred to, is most notably going to create a rather twilight zone existence. Companies are going to be able to track things that they have never had access to before. For instance, how often did you eat your toaster strudels in the morning? You may be able to let the company know how often you buy it, but they don't know exactly how often you eat it. They may know that you buy it every once in a while, but they don't know how often you actually sit down and eat it. That is a problem. They will now have access to that kind of information. The microwave will have a profile built exclusively to cook toaster strudels. It will then keep track of every time you use the microwave or even the toaster to make your toaster strudels. Companies will now have data on how often you use products not just how often you buy them.

## The World of Automation and Future Laws

Eventually, everything will be automated. This also includes the formation and implementation of laws. It's similar to that of an episode out of Star Trek where the Borg attempted to assimilate everything. You essentially won't need to work because the work is already being done by robots. In fact, the only potentially secure job in the future is robot repair and there will be subclasses of repairing. There will be robots that do the general number of repairs, but highly specialized repairs might still be done by humans.

This brings up a very interesting dilemma and that is your place in such a society. Artificial intelligence is a long way off from being general intelligence, which would be capable of running itself. However, we will get there eventually and many of the highest academic professionals fear such a time. After all, the human body is full of flaws and problems. We get cavities, heart attacks, sleep, and many other things. While we are nowhere near a solution to such a problem, the problem edges closer every single day. There are already artificially intelligent machines that run core components of our lives. I don't know of a single person that could live without Google Maps when going to another city. Additionally, how would one find minute flaws in a nuclear reactor or rare cancers in a human body without AI at a near perfect rate? Artificial intelligence runs a lot of background gears in our world clock and many of us don't even know it. Artificial intelligence only works when given good statistics.

In this book, we will explore how statistics are used in almost every sphere of life; how statistics can be used to create positive change; and also how accurate statistics can be used in a misleading way to benefit the few and cause great harm to society at large. We will also look at how well-intentioned use of statistics can sometimes cause

damage because of carelessness in selecting the right type of statistic.

# Criminology and Law

## The Potentially Pregnant Model

*Incident*

We all experience controversy every now and then but there are few examples in the history of statistics that show us why we need to be careful. Imagine that you are a young woman that is in high school and your parents keep getting coupons for baby items. These coupons are rather new and while they are sprinkled in with other discounts, they are quite noticeable to your parents. Your parents take offense to the fact that the store keeps sending these coupons even though it's useless to your family.

Therefore, your father goes into the store and complains about the fact that these coupons keep showing up. The store owner knows nothing about this then apologizes and says they'll try to fix the issue. A couple of weeks go by and you eventually succumb to the pressure of telling your parents that you are pregnant. Your father is of course

15

flabbergasted at the fact that you are pregnant, but your father accepts it for what it is.

The manager attempts to contact the father to see if the issue has been resolved but the conversation goes a little bit different this time. Your father apologizes to the manager and tells the manager that the coupons were actually needed. The manager is just as shocked as the father is but still has no idea how the store managed to send the right coupons at the right time.

The reality of the situation is that a statistician was hired to potentially exploit pregnant women. Items concerning babies are very expensive, even though it's not usually that expensive to produce the items themselves. I mean, how much is a diaper in the cost of materials? How much is 4 tablespoons of peas? Yet a single diaper could cost $3 and a baby jar food of peas usually costs about $2. Needless to say, it's a good market to try and make money in.

Therefore, the statistician was hired to try and figure out if customers were pregnant ahead of time. The statistician looked at all of the women who later had babies and compared their purchases. This led him to find out that they all tended to buy very similar items months before they actually had the baby. In fact, thanks to this information, he was able to not only predict which women were pregnant but how far they were into the pregnancy.

In order to cover up the fact that they had figured out a formula to predict whether women were pregnant before some of them even knew, they couldn't just hand over coupons. It would be a little insensitive and jarring for a store to suddenly send you pregnancy-related coupons. During the first batch, you might not have any questions but every batch after that might seem a little suspicious to you. Therefore, in order to hide the fact that they had this formula, they began to sprinkle these items along with other items that were completely benign. This would make the coupons seem more natural and generalized. They wanted to do this because they didn't want any potential backlash for the algorithm that they had discovered.

The problem is that humans are not predictable creatures in standalone environments. There's always going to be something that doesn't go as planned. The story that I just shared with you was a real augmented story concerning Target. They had figured this formula out and the actions were pretty close to what I shared with you, but they ultimately arrived at the wrong outcome.

*Outcome*

There are two different outcomes to any statistic that misleads the public. The first outcome is usually that the company gets what they want. This is because people are often too lazy to read statistics or simply don't know about it. In fact, given the situation here, one could argue that no one really knew except for the company. The company was extremely successful in making an algorithm that detected whether women were pregnant before some of them even knew.

The problem is that not everyone could see how this was misleading. After all, it was found out that the statistic was used to advertise coupons to young women that might be pregnant. The part about it being misleading is they were getting coupons based on general likes. After all, people don't really view grocery stores as being the next version of Big Brother. Target wanted to sell a specific product to a specific audience and used data to achieve it. The problem is that most people don't like being tracked by their purchases, but readily accept those little discount cards that allow them to get coupons.

This leads to the second outcome that is usually very different from the first outcome. Once the public finds out about the issue, it becomes a different situation. Almost everyone thinks that coupons are generalized and not focused on the individual. They understand how Amazon could be personalized, but many of them don't even understand that places like Target do the same thing. Your credit card usually doesn't change and that means that they can actually keep track of what you buy on that credit card. The moment that you use a coupon, that coupon has a code on it. That code corresponds to your mailing address because it went out in a specific book. That allows these companies to associate mailing addresses to credit cards.

That's provided you actually get a Target booklet of coupons in the mail. Most people fall victim when they sign up to the website and use their credit card there. That gives them the ability to send you coupons over the internet and at your mailing address.

However, regardless of how you get your coupons, there is a boundary companies have learned not to cross. Yes, it was effective in predicting pregnancies, but it was also effective in invading personal privacy. While they do continue to do something similar to it, in order to avoid a public backlash, they do their best to avoid crossing personal boundaries. Public boundaries are your tastes in music, clothes, and similar things. Personal boundaries are things like having a baby, premature ejaculation, and that list goes on and is uncomfortable.

It has been proven with gay cruises, pregnant teenagers, and drug addictions truly personalized ads are not a marketing solution. While they do work for a short period of time, they do cause quite a bit of problems and confusion for the people receiving them.

*The Concept That Proves a Wrong Outcome*

The type of study that they did is actually a fairly common one. All they did was keep track of the average purchases of the customer. That is actually something nearly every company does regardless of circumstance. However, this is what's known as a statistical average. It's actually very easy to calculate because it kind of goes along with what we think of as common sense.

If you are customer that comes in every single day and gets a cup of coffee of Colombian Brew that is at 120 degrees Fahrenheit, it's predictable. It means that we can assume that you will continue to do so unless something drastic changes in your life.

Having said that, this study was a lot more detailed than my own short analogy. The statistician would keep track of things like lotion, baby validators, specific types of food, and the list goes on. So, you might be wondering, what happens if people continue to buy things like that but also buy other things? This is when you get into statistical probability. Let's say that we have 100 female customers. 10 of those customers buy lotion. Another 10 of those customers buy lotion and tampons. Another five of those customers buy lotion, tampons and birth control.

Therefore, all together 25 customers buy lotion, 15 buy tampons, and 5 buy birth control. If a woman is buying tampons, the odds of her being pregnant are statistically low due to the fact That pregnancy usually prevents menstrual cycles from occurring. Knowing this simple fact, one can calculate the probability that if a customer doesn't buy tampons for 2 months in a row, it's highly likely that this customer is pregnant. Now, you also have the problem of male vs female being in that equation. I'm not going to get down to the nitty-gritty mathematics as this is a fundamentals book, meant to introduce you to the concepts behind statistics. However, sometimes you have extraneous variables that seem related.

The statistician that worked on this formula also had to deal with this. What the statistician did was figure out the average combination of products that were exhibited in customers that had children. It's actually really simple when you think about it, but the application of such math is far more difficult.

## Old Lady's Stolen Purse

*Incident*

The standard story of being mugged while out and about happened to an old lady. The only thing that she could remember was that the female had blonde hair and was white. The male was white as well with brown hair and a beard that went down below his neck. They then got into a car and drove away.

Now, the average person might look at this and go "Well if that's all she remembers then she's not likely to catch the criminals". However, thanks to a statistician, that story is a little bit different. Sometime later, the cops caught two individuals that match the descriptions provided by the victim. In order to provide evidence against these two individuals, the cops brought in a statistician that said it would be 1 in 12 million possibility that it was someone else. How did he do it?

When it comes to statistics it's actually quite easy to figure out how he calculated the probability in which it could be someone else. He had a few things to go on. Needless to say, the two individuals had to be together at the exact same time. The female had to be white and had to be blonde. There is a percentage in which the town population would be white. There is also a percentage of the number of white individuals that are also blonde. Just like the female, there's only a certain amount of people in the city that are white males. In a subsection of that, there are white males that have brown hair. In a further subsection of that, there are white males that have brown hair but also a beard. Finally, even further into that subsection, there is only a small handful of those white males with brown hair with beards that have long beards.

This means that only a certain handful of people could have possibly done this. Beyond that, the two were a couple. Meaning, it is highly likely they would be together when this happened. This is what's known as refining an average. There is an average for how many females to males there are in a single city. The natural hair color of an individual can be brown, blonde, red, and black with varying shades. Yes, you do have the off chance that they dyed their hair but that would be ruled out by asking for the natural color. This means that the female would be a certain percentage of society and then of those females, ¼ of ½ of them would be blonde. This would come out to ⅛. A female can either be in a relationship or not be in a relationship, which means it's ½ and the total goes to 1/16.

On the male side, you can either have a beard or no beard. This means you already have ½ portion from being male. With the color of hair, it goes to ⅛. With the beard it goes to 1/16. If you have a beard, you can have varying lengths of the beard. Beard lengths can be considered stubble, short, medium, or long. This brings it up to 1/64.

Now for some additional information that was hidden. A mugging can happen with or without a weapon. That's ½ applied to both male and female, so 1/32 for female and 1/128 for male. You can either be shorter or taller than the victim, which means another ½; 1/64 for female and 1/256 for male. You can either have a car, a bicycle, or not, so ⅓; 1/192 for female and 1/768 for male. They can either be at work or not, which is another ½; 1/384 for female and 1/1536. Finally, it can be day time or night time, which is ½; 1/768 for female and 1/3072. 1/2359296 is the answer I get for the probability that it was this couple for when you combine the two.

This long series of mathematics was to show you how probability is calculated whenever you are simply looking at distinct features. If the population is less than 2 million, I'm pretty sure they're guilty. In fact, when told that it would be a 1 in 12 million difference for it to be someone else, the jury came out with a guilty verdict.

*The Concept That Proves a Wrong Outcome*

The problem with this statistic is it assumes everything is in stone. There is a set number of blonde people in the city and we somehow know that. There is a set number of females in the city and we somehow know that. Everything has a concrete number to it, which is not true of a city at all. Cities have homeless individuals, people change their hair color from brown to yellow, and really the only reason why they got away with this was because it was in the past.

Statistics are good at proving the likelihood of something occurring. Understanding that I came up with my own version of the probability of it being these two should show you the varying concerns of statistics. Yes, statistics are very good at creating a probability matrix that tells you the likelihood of an event. However, that doesn't mean that the probability matrix is right. A probability matrix is affected by unknown variables, unforeseen circumstances, and pretty much anything that's just generally not known. If the female had a birth defect, it would definitely reduce it but if she didn't live in the city then the entire model is useless. The entire model is useless due to the fact that the statistician's information was based on the population of the city.

While it is definitely convenient to bring up the probability of it being a different combination, it's not realistic. There are several factors that went unaccounted for and it's circumstantial evidence at best. It's saying that given the circumstance that they are residents of the city, this is the probability at which it is someone else. The problem is that muggers have been known to go to other cities to mug people because that makes them much harder to catch. It actually makes them much harder to catch for this very reason.

## Mass Incarceration: Repairing Past Damage caused by Statistics

Now the problem with the law being ran by statistics is that a lot of it can go wrong. It can go wrong because of people who build a statistic while having underlying issues. Underlying issues like racism and being anti-poor. You could use a statistic like saying that all colors of a certain race were twice more violent given these circumstances, which means we need to make a new law about it. Equally as likely, you could say that the filthy streets are the cause of the poor on the side of the road and provide a statistical analysis that proves that without the poor, we would have cleaner cities. You could use that statistic to create a new law that banned being poor and if you were poor, you would have to be removed from the streets.

On top of this, the average person only cares about what's relevant in their lives. Because there's so much going on in the world sometimes it's literally impossible to know what's going on everywhere. It was hard enough back when we didn't have the internet to understand what was going on in your country by itself. Now that we have the internet, we get to know when something bad has happened in France, something horrible in Syria, or something fantastic happened in India. If it wasn't for people who were dedicated to covering a single nation's general issues some of us wouldn't even know about those issues in the first place. Do you actively try to contact local news places in another country to find out what the latest news is? That's the point. Ill-intentioned politicians use statistics cover up bad laws. In America, we are still repairing the bad laws created from the drug war. As of writing this book, tens if not hundreds of people are being affected by the same laws created during an era in which people just didn't pay attention to the statistics behind drugs. They just knew that the situation was bad, and politicians seemed to have a good answer based off of numbers. We now know, they weren't really concerned about drugs in the first place.

In addition to this, according to a couple of conspiracies, they may have even started this drug war. Whether that's true or not, I cannot say but the point is that laws were put into place based on bad statistics and that's why it's important to care about how you get your statistics.

# Business and Marketing

## Colgate study

### 80% of Dentists Approve of Colgate

The Colgate study is an interesting example of when a company doesn't technically lie. The study claimed that 80% of the dentists that Colgate surveyed suggested their approval of Colgate as a toothpaste. The reason why this wasn't technically a lie is because they did survey dentists, they did ask them their preferred toothpaste, and 80% of them did reply with Colgate. However, the dentists had the option of providing multiple toothpastes that they would suggest.

As you can probably already tell, this meant that Colgate was suggested but also with other products. However, the customer didn't need to know that, and it wouldn't help Colgate increase its sales. Instead, Colgate took the fact that 80% of the doctors did list Colgate and just never informed the customer that that was how the study was done.

Having said that, once the advertising standards control authority figured out what happened, they ordered Colgate to take it down. They knew that the average customer would not be able to understand how the study was performed. In fact, they generally theorized that the average consumer wouldn't even bother looking at the study. Instead, they assumed that the customer would think that it was a one-answer survey.

However, because Colgate didn't technically lie, the company didn't get into a lot of trouble. In fact, even though it had been taken down, it wasn't exactly published very much, and customers already had the imprint. Customers had already seen that a large percentage of dentist that approved of Colgate and the damage had already been done. At that point, Colgate had effectively advertised to these customers before standards in place stopped them. Stopping them just prevented them from getting even more customers.

## What the Customer Sees

The advertising control authority understands that your average consumer is rather lazy when it comes to consuming information. If we tried to research every claim that every company ever makes, we'd never have any free time. Saying that this car is the safest for this year would have to involve researching every car and their impact on safety. Try doing this for the 20% more bottles on shampoo or food. As you can see, trying to figure out if a company is lying to you through statistics is quite a hassle for the average consumer so it just isn't done.

Instead, what actually happens is your average consumer sees the headline and just assumes it's either true or not. By having a big, well-known company claim something as a fact, many consumers just believe it. This is often known as authority bias and it happens with those in seeming positions of authority. Colgate is a very large toothpaste brand and so it kind of makes sense for dentists to recommend them. They wouldn't be as large as they are if they weren't effective at doing their job, right?

Well not necessarily. You don't need to have a product that's good at its job to make money off of that product. You just need to convince the customer that the product is as good as you say it is. Once you convince the customer of this, they will believe it and then something special happens when they buy it. Due to the fact that they bought it, they have a bias to believe that it is better than it actually is. This is kind of like Starbucks Coffee, which is really not that special. I mean, it's coffee and we have had coffee for years. We have had all sorts of different cafe shops where you could buy specialized coffee. However, with the right branding and marketing, they can convince you that they have delicious coffee at a convenience and *they were better*. Mind you, that convenience costs a lot in money, caffeine consumption, and sugar consumption. Once you buy that coffee, that coffee will actually taste better to you than your own coffee because it's a choice you made. That's not always the case but it happens more often than not.

## The Reality

This brings us to the reality of the situation and that is that companies will do their best to convince their customers to buy their product. Customers are often too lazy to double-check the work that these companies did. Companies understand that their consumers are often too lazy to check the work that's been done. Therefore, companies regularly take advantage of lazy consumers and advertise to them from that vantage point.

The standards put in place for marketing are often after the fact measures that can be taken against the company. It's actually kind of ironic that there isn't a position of authority where companies are required to submit their advertisement for approval. We have tons of methods for controlling food, drugs, and even equipment but not advertising. There are several unsavory methods of advertising that companies use and have used over the years.

The most disgusting in my opinion would be that of advertising to children, who are highly suggestive individuals. There have been steps against such advertising and limitations placed on it, but I still think it's somewhat disgusting. Companies understand that children will look at these advertisements and then ask their parents to purchase the product. Therefore, all the companies have to do is convince children that it's something they want. The children then go through the process of convincing their parents to purchase the product. This is why the toy aisle inside of large grocery stores is almost half the size of the grocery part of the store. The stores understand this method of advertisement and actually assist it because they make money too.

These two examples just bring to the forefront that companies will attempt to do whatever it takes to convince someone in your household to buy a product. Whether it is through the statistics that you're too lazy to look further into, the children sitting in front of their favorite kids show, or even just manipulating the colors you see. These companies only follow the law when it comes to marketing not moral values.

# The Sampling Bias in Marketing Claims

## 9 in 10 [Insert Here] Recommend [Insert Product]

We've practically all heard the commercials that say that a certain percentage of professionals (within a relevant industry) have given their thoughts on the product. The problem is not that this is a lie, it is a controlled sampling issue. The doctors that they measured were hired by them and had favorable outcome. In other words, they bought favorable responses so that they could claim (legally) that something was recommended 9 out of 10 by specialists.

This actually happens quite a bit because it's very easy to employ as a tactic. There are a few benefits to having such a claim on your product. The first benefit is that the specialist is often trusted to do their job and usually has no reason to not do their job. The second is that most people trust these specialists to do their jobs impartially and so they have a sense of authority. Finally, by paying these doctors or specialists for the recommendations, you can legally say that you obtained it. This way you can make a grand marketing claim without having to worry about any backlash. This is a dead on example of sampling bias committed by marketing agencies.

**[Insert number]% More [Insert Item] (small text of *average comparable item*)**

This one is a bit more simplistic than the previous one because all they need to do is choose a popular product. You see, they don't need to survey every single product on the market to make sure it's true. All they have to do is make sure that the product they are comparing it to is rather popular.

For instance, let's say that we have a loaf of bread that we wants to claim has 20% more fiber than your average competitor item. We put a little asterisk next to it for legality reasons and then specify we compared it to products that have at least 20% market share. Let's say that one of those loaf of bread competitors is Wonder Bread. They have 2 grams of fiber inside of their white bread. We didn't say that our wheat bread has 20% more than Wonder Bread wheat bread, we just said that it has 20% more than your average bread made by Wonder

Bread. If we have wheat bread that is 4 grams of wheat fiber and Wonder Bread has an average of 2 grams, we are legally in the clear (given the serving size is 10 grams). Meanwhile, someone who actually has 20% more than your average wheat bread provider will have somewhere around 6 or 8 grams of fiber. However, our marketing team can save us money by using technicalities and since almost no one checks the asterisk then we generally get an increase in sales. It's easier to compare your product against Wonder Bread, than compare it against the entire market.

As one can see, the average person would consider this to be deceitful or lying but marketing companies don't view it that way. As long as they can't get sued for it, marketing companies will straight up lie to your face if they think it will help you buy a product. This is why we have laws in place to make it much harder for marketing companies to lie to your face and why they use technicalities. With technicalities, they don't have to be lying but they can bend it to where it's almost identical to lying.

## How Businesses make "est" Claims

### Words are Subjective but Seen as Objective

Many people like to speak objectively. This is known as nuance because we don't really deal in objective speech. The average person might say that their smartphones is really great. In fact, they might even claim that their smartphone is the best.

The funny thing is that it is this mentality that allows customers to see messages from ad agencies that say their phone is the best. Remember, all they have to do is put a little asterisk next to it saying it's the best given certain circumstances. The law is a subjective but objective word that everyone agrees to follow. It's subjective because the law was created underneath a certain set of circumstances. Laws are further refined by more circumstances that come after that loss has been made. However, that law has an objective outcome.

Knowing that the law has an objective outcome, companies can then use this to undermine the law. In fact, they only ever really get forced to do something if the customers don't like how the company undermines the law. Then those customers bring in a class lawsuit and the law is further refined to remove that way of undermining the law in the future. Therefore, when a company says that their product is the best they are usually using very strict parameters in which that's true.

## It's the Best, Fastest or Strongest *given this*

Therefore, let's explore an example. Let's say that you're going to go buy some toothpaste and that toothpaste claims that it is the best toothpaste on the market. You check the back of that toothpaste and it might say the claim to be the best was made after averaging all natural competitors. To you, that might seem like they are talking about just their competitors. They looked at all their competitors and objectively found that they were better. However, the key word there is natural. There are all natural-based toothpastes out there that have statistics on how well they prevent cavities. A regular toothpaste can then claim they were the best toothpaste based on its success rate versus those of natural toothpaste. You might think that it is not a fair fight, but the marketing company doesn't care.

This isn't to say that all companies do it because sometimes they really are the best. However, most companies do it because you have to use a metric in which to compare yourself to find if you're the best. Therefore, using a less competitive market to you is more ideal in that situation if you know your average customer is not going to pay attention.

## Companies Often try to Increase Sales through Clever Wording

Have you ever heard that Macs cannot get viruses? A lot of people still believe this, but it's not actually true. However, if you noticed, Apple really never corrected the statement. Companies will use clever wording and statistics to convince consumers to buy their products. Most of the time they will use half-truths to get the job done.

A half-truth is where they present partially true material but then use false statements to represent it or what we would *subjectively* consider false. Therefore, back when PC was trying its best to fight against the Apple machine, it would often claim it was better for work and enterprise. Meanwhile, Apple's claim said it was for the creatives and both sides would give subjectively objective reasons. Both sides of this were actually doing half-truths. I'm not sure if you noticed this or not, but Photoshop was just recently placed on Apple products, like within the last decade. Adobe products started out on the PC, which means that one of the most creative suites on the market was on PC.

So, if that was really true about what Apple was saying, Adobe should never have been an industry-standard. It should never have been the software that most artistic students use. Yet Apple looked at how their average customer described themselves and claimed that was the person or type of person that use their products. It's not 100% true and it's not 100% a lie, therefore they don't have to worry about being sued over it. A lot of companies do this because they can get away with it. There's no law that prevents them from doing it until the law is created and then they move on to the next place.

## Use of Small Sample Sizes by Marketing

## Online Twitter Poles

Twitter is an interesting example of how one could gain a biased sampling. Twitter is a very large platform that has a lot of people attached to it, but the problem is that a vast majority of the people on Twitter think a certain way. In fact, one can generally see that when companies do online polls on Twitter, they generally know exactly what kind of results that they are going to get.

Twitter has placed itself in a rather interesting position because Twitter represents something for the 1990s with 2019 logic. You see, Twitter is what's known as an open forum which is to say that pretty much anybody that is signed into the platform can comment and see practically everything. That being said, they have upgraded their services to provide privacy. In the 1990s, it was actually kind of common to see a forum online where one could participate in a discussion. It was fairly common because it was fairly easy to build and there were plenty of services that would allow you to build your own forum.

However, once the smartphone was built, the online forum seemed to deteriorate. It's not really particularly clear as to why the online world did this. I suppose the character limit on Twitter allowed for online forms to gain spotlight attention in another way.

The thing about forums is they often had rules as does Twitter, but Twitter has some very biased rules. Twitter bans so called "hate speech" on its platform, which is a very left-wing term. In fact, it's so left-wing that the right-wing tends to call it "wrong speech" or "wrong speak" because it doesn't make sense why it is wrong, but it's commonly known to disagree with left wing opinion.

## Neighborhood Opinions

Another trick that marketing teams love to do is to grab the opinions of the average buyer. The only problem is that the average buyer doesn't really want to give their opinion. You might think that this is a bit of a conundrum, but you see they have their own "average buyer". Their average buyer says whatever they want them to say because they got compensated or already like their company.

Actually, in many places it's illegal to compensate for positive reviews. Therefore, companies do selective reviews. On your average website you would see this as glowing remarks from the latest customers, only they're not really the latest. In fact, you don't know they're the latest because they usually don't include a date on which the review was made.

So, they attempt to send out surveys to customers that liked what they did. They don't really send out surveys to a lot of customers that had a problem. In fact, many of these marketing agencies actually keep track of the "problems" so that they almost never send something out to them. It has actually evolved into sending out a survey to everyone but categorizing them with AI so as to filter out "inconsistent data".

Therefore, when you see statistics that say that their average customer fares better than another, it's not really a lie but missing information. This is something they love to do because it's not technically lying and so thus it is not illegal. They also, to skirt around the law, include some of the problems. They will sprinkle in a few of their problem customers to make the statistic more realistic. This allows the average customer that doesn't know any better to believe them, but anyone who has a certain level of critical thinking knows better. However, that type of person usually isn't the company's go to for a customer. In fact, many companies don't want customers that can think on the level they can.

Look at Starbucks. What exactly do you get from Starbucks? You get tasty coffee? If I were to make what they make, it would cost far less. Not saying this is the official recipe, but this is how I make a caramel macchiato. I place about a tablespoon of maple syrup on the bottom, 2 shots of espresso, and then fill it up with milk. I wait until I get to about 10% room left. Then, I use a steamer to rise some of the milk up into a foam. Finally, I use heavy whipped cream on top and drizzle some caramel. In combination, the ingredients usually cost me less than $20. I can also usually make about 20 to 30 caramel macchiatos this way. The average Venti Caramel Macchiato that I get from Starbucks costs me $8. If we did the math, we would see that if I went to Starbucks every time I wanted one, I would spend $160 to $240 versus $20. In other words, they make $140 to $220 of profit on top of just me.

They don't exactly want you to understand how much money you are spending for the simple convenience of buying a cup of coffee. They want their brand associated with top quality coffee that tastes great for a decent price. Only, that decent price is not really decent to the average customer once they figure out the cost difference. Therefore, when they do surveys on how customers like their services, they will often go for the higher end purchasers. People with tons of money to continue purchasing Starbuck products without ever needing to worry about cost. These people are far more likely to give them the results that they want so they can put it up in advertisements or in flyers.

## Small Samples used for Marketing

This actually brings me to another point that they like to do, which is to say that they like to take small samples. Just as with the surveys that I mentioned, they often take small samples of preferred neighborhoods and people. Therefore, they're not likely to survey the people out in Decatur, Illinois but rather New York, New York City. This is because the population of New York City is much more inclined to be okay with the high prices of Starbucks coffee. Decatur, Illinois is a relatively small town in comparison with a drastically lower budget for overpriced coffee. Therefore, they're much more likely to receive glowing reviews from the city that's okay with $8 for a cup of coffee.

While I may be harping on Starbucks, it's not the only company that's capable of doing this. In fact, they likely have never done this because I don't think that they have ever needed statistics, I was just using them as a potential example. Companies that collect statistics that will allow them to promote their business will tend to choose small sample sizes in neighborhoods they understand. Therefore, if they choose a low income area to judge their high-priced products, they're not going to get the result they want. Instead, they would rather choose a high-income area that matches their high-priced products to justify saying 9 out of 10 customers love their products.

This is known as small sampling and any morally just statistician trying to do the job of a statistician would never do a small sample. In fact, when you have a population the size of New York, the statistician would attempt to collect data from as many people as possible. Usually, the survey would contain somewhere around 2% to 3% of the population of New York. Companies don't work like morally just statisticians though as they want to look as good as they possibly can.

The reason why a morally just statistician samples a population is because the population is too large to do by oneself. If we took the population of New York City and multiplied it by 5 minutes, the time to take survey, you would have how long it would take. However, statisticians discovered that by looking at larger groups of data, you could sample a certain amount of a population and gain an average.

With ten people, you may have five different opinions. With a hundred people, you may still have five different opinions. With a thousand people, you may still have five different opinions. What they found was that if you increased the amount of people you survey, you are more likely to get the same answers. This is because humans are predictable and habitual. Therefore, a morally just statistician will collect a sample of the current population that's decently sized, preferably above at least 1%. This is to prevent small sampling, where you have very biased results based on where you do it.

Notice I also said that the statistician would measure all of New York not just New York City. That wasn't the shortening that happens with Americans that just refer to it as New York. A morally just statistician would attempt to do the entire state to remove the potential wealth bias that New York City has versus everyone else in the state. A company doesn't want to do this because then you end up with a similar result to Decatur, Illinois. Therefore, they will do it in the highest income area where they have a good portion of customers that are very satisfied with their product. This is an example of how some companies use small sample sizes from known neighborhoods to get statistics that skew in their favor heavily.

## Poorly Chosen Averages

### The Average Living Wage

When it comes to poorly chosen averages, none represent them best as the average living wage. The average living wage is a statistical myth that politicians push practically everywhere. The truth of the matter is that you could not just base the living wage off of the numbers in the area.

The average person would think that a statistic like the average living wage shouldn't be a problem. The problem lies in the average part. Someone in New York City is not going to have the same living wage standard as somebody living somewhere in North Carolina. New York City itself is an extremely expensive place to live in and many of the working-class individuals actually don't live in the city. Instead, they live in the areas around the city or they got lucky and locked in a rate long before it got crazy. In North Carolina, no offense but you don't really hear anything about rent in North Carolina. I specifically chose North Carolina because it's rather unnoticeable as a state. I mean, do you know of any celebrities that live there or national monuments maybe? Needless to say, it's got a much lower living wage than those living in New York City.

That isn't to say that there isn't anything wrong with North Carolina, it's just pointing out the difference in scale. Therefore, if you talk about the national average living wage, it's a completely unrealistic number. You are legitimately taking the most expensive cities and the poorest cities to make an average number that doesn't work for anyone. Now, let's get a little bit more micromanaged here. Let's say we're just talking about New York City itself. Even then you have a problem because there is the rich side of New York City and the poor side. Given the number of millionaires and billionaires currently residing at New York City, you can see how skewed such an average might be.

All right fine, let's divide it between the wealthy and the poor. Let's see what happens when we try to take an average living wage of the wealthy and the poor. Well, how wealthy is wealthy? How poor is poor? These are not easily defined. If you define wealthy as anything above a million dollars, there are quite a few people who have above a million dollars in New York City. Meanwhile, what is the definition of poor? A person living from paycheck to paycheck? A person who doesn't even get a paycheck? A person who doesn't live in a housing construct?

As you can see, trying to conceive of an average living wage is veritably impossible due to the different circumstances that everyone lives in. Yet, the way that we assist people is based off of this metric. The way that we make decisions to improve people's lives is primarily based on this statistic alone. If this statistic has such a problem, why are we basing anything off of it? This is a poorly chosen average that continues to persist even though the gap between what is wealthy and what is poor is never going to stop growing apart.

## One-size-fits-all Clothing

I've heard countless stories about how the one-size-fits-all doesn't ever actually fit everyone. The one-size-fits-all clothing is based off of national average weight and height. The national average of height usually runs around 5ft 9in for males and 5 ft 3 in for females. Meanwhile, the national average of weight usually runs around 120 lbs for females to 160 lbs for males. Therefore, if you are 6 ft 3 in and weigh in at around 250 lbs, there's no one-size-fits-all clothing that will fit you. On the other hand, if you are 5 ft 1 in and weigh 90 lbs, it's not going to fit you either.

This is a type of clothing that serves as a dual purpose, but no one ever pays attention to the second purpose. No one pays attention to the fact that the design of the clothing is meant to shame those that are not within the average. I'm not saying that these companies specifically set out to design their clothing like that, I'm just saying that it is what it is. If you don't fit into the average, this clothing will not fit you and therefore you will feel bad by not being able to fit. The natural response is to feel as though you are not normal.

You might wonder why such a metric is important. Well, considering the vast number of companies and corporations that pour enormous amounts of money into employee clothing, it's a bit of an issue. Companies that try to save money by creating a one-size-fits-all end up excluding anyone who isn't average.

Have you ever seen what's known as a Farm Boy? A farmer boy is referred to as such because their body is built a specific way. They are usually very tall and quite heavy. This is because on the farm it is beneficial to have room for muscle mass. After all, for a very long-time farmers had to carry in barrels of hay and corn on their backs before wagons. Even when they had wagons, they had to pull the wagons until they could get a horse to pull it. This led to a specific body type amongst Farmers where they were heavy in muscle and very tall.

In the exact opposite direction, you have what's known as a City Boy. A city boy is usually a lot smaller and a lot skinnier, but more importantly is usually the reference model for the average. It is the city boy that the average one-size-fits-all fits because there are more City Boys then there are Farm Boys.

Therefore, how can you create clothing that is supposed to fit everyone yet is only built for a specific body type? While not as impactful as one would want an example to be, it shows a perfect example of a poorly chosen average. In fact, it represents a poorly-executed product. You see, one-size-fits-all things was a very popular product in the 1990s and in the early 2000s. Nowadays, you would be very lucky to see any clothing advertising that it was a one-size-fits-all because those companies tend to lose money on it.

**High GDP Means Great Country**

GDP is often a measurement used to measure the success of societies around the world. We have several different countries and people often want to quantify their success. Therefore, the national GDP has been a long-time staple for many in the industry. However, it's honestly a very poorly created average to represent us. GDP stands for gross domestic product, which is to say how much product is moved in and outside of a specific country. Therefore, if you have a high GDP you are likely to have a very high threshold for making money in that country.

This is very important if you want to create a company in that nation. It's very important if you want to move to a country where your currency is worth more than that country's currency. It has its importance, but it doesn't measure the level of quality of the country itself.

Let's say that you go into a store and you are greeted by two different employees trying to sell you something. One employee is dressed in a very nice suit and tie and talks in a very clear manner. Another employee is dressed in drabs and has an accent you've never heard before that makes it very difficult to understand them. Furthermore, you think that they might be insulting you at nearly every chance they get. Which one would you go with?

If you said you would go with the one that had the nice suit, you would have gone with the person from Ireland. Ireland has a GDP of

379 billion $ as of writing this book. Additionally, as of writing this book, The United States of America has a GDP of 21,482. What about the person in the drabs? What about the person with a heavy accent you could barely understand? What is that person from?

That person was maybe from Saudi Arabia or Nigeria or the United Arab Emirates. Saudi Arabia has a GDP of 795 as of writing this book. Nigeria has a GDP of 447 as of writing this book. The United Arab Emirates has a GDP of 445 as of writing this book. Now I'm not saying that these are iconic images from those countries. I'm not saying that everyone from those countries are in drabs. What I am saying is that these are countries the average person would not normally want to visit.

Saudi Arabia has quite the unique law system that is particularly friendly towards killing criminals. Nigeria is kind of already known for its fake princes. These are countries that are not really that friendly when it comes to foreign travelers. However, their GDP is really high. If we solely chose what we were comfortable with, some of the highest

GDP countries would never see us. This is because some of those countries have human rights issues or criminal issues.

GDP simply represents how much a country can make on average. This average is often used as a justification for propping up countries.

The problem is that some of these countries are countries that foreigners would typically stay away from even though they have a high GDP. For instance, a good reason why Saudi Arabia has a high GDP is due to their oil reserves. They provide a lot of natural gas that the world uses and so it's kind of a must that the world trades with Saudi Arabia. Yet, a normal Christian white person would be at least reluctant to step foot in that nation.

**Poorly Chosen Averages Mislead from the Truth**

The reason why poorly chosen averages are important to notice is because they mislead you away from important issues. I mentioned Saudi Arabia because it's a country known to have a past with human right issues. I talk about Starbucks because of the exorbitant price they get from their customers. I talked about how companies will use every trick in the book not to lie to you but to trick you into believing they are the best. These are all examples of where the averages are skewed.

Let's talk. It's a little bit more serious than that. A pharmacy company has produced a new drug for curing cancer. Now according to the study in the research, it cures cancer 99% of the time and it's for every type of cancer there is. A relative of yours is a victim of cancer and has to take the drug. The drug costs $1,000 per pill and your friend only has enough for one treatment. One full round treatment of these pills. It's going to cost them $30,000.

They take the treatment and sure enough the pills do indeed cure the cancer. However, when your friend goes into the primary care provider next year, they've got something new. They've got a ramp in case of tuberculosis that's very aggressive. The doctor says that it sprang up out of nowhere and that they don't really know what caused it. The pharmacy that produced the pill created an average of risk to the user. In other words, when a pill goes through the Drug Administration for approval, a risk assessment is given to each pill. The company is supposed to tell you what else the drug might do to you.

However, what that company chooses to tell you is solely up to them. In addition to this, the doctor has the ability to tell patients what might happen to them because they took a drug as well. Well, in the fine print of the very large instructional paper that came with the pills it says some users may develop tuberculosis. This is a horribly placed average because, in some cases, tuberculosis is worse than cancer. Yet the company is only required to tell you there might be a risk, so they weren't lying about it in a way. Even though it says 20% in their statistical data, they can say some. They can reduce a statistic that says that one-fifth of all patients who take this medicine will develop tuberculosis into a descriptive word like *some*. Since *some* represents an undefined quantity larger than one or two, it technically fits. In addition to this, because the company warned you ahead of time, you can't sue them to try and get money to cure or at least help your tuberculosis.

I took that to its extreme because it's very important to understand the motives and methods used by academically corrupt statisticians. A pharmacy doesn't want you to really know about the risks of your pills so long as it gets the job done. I'm not saying that all pharmacists do their best to make sure you don't know the truth. What I'm saying is that pill makers are among those that play down the importance of "harmful to their business" statistics.

# A Message from the Author

Hey, are you enjoying the book? I'd love to hear your thoughts!

Many readers do not know how hard reviews are to come by, and how much they help an author.

I would be incredibly thankful if you could take just 60 seconds to write a brief review on Amazon, even if it's just a few sentences!

Please head to the product page, and leave a review as shown below.

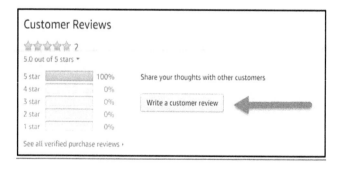

Thank you for taking the time to share your thoughts!

Your review will genuinely make a difference for me and help gain exposure for my work.

# Social Sciences

## 2 True Percentages

Does a 1.0% increase equal the same thing as 100%? You might think that this question is rather odd, but in statistics it's quite common. Let's say that we are talking about the dropout rates of individuals going to med school. Last year, 5 students per 1,000 students dropped out of the program. This represents 0.5% of the school receiving dropouts per 1000 students. Let's say that this year, 10 students per 1,000 students dropped out of the program. In such a scenario, we would say that this is 1.0%. Therefore, we can say that this is a 0.5% increase, but we can also do something else. Since 0.5% doubled is an increase of 100%, as it is increasing by its own number, we could also say that the rate of dropouts raised 100%.

In such a case, you have two different percentages that equally represent what the data is trying to say. You have doubled in dropout rates since the study done last year. However, you have also only experienced a very minor 1% dropout rate. That is actually an incredibly small dropout rate considering the average in the medical industry. That's not what we are here to talk about though, because the idea that two numbers represent the same thing is foreign to us.

On one hand, we have a statistic that shows our data represents a very small portion of the population we are taking it from. I mean, 1% is pretty low. However, on the other hand, we are also talking about a statistic that shows a huge increase. In fact, if you charted it, it would still look small. If you recall, companies try to utilize the small vs. big visual representation to make their company look better. Therefore, you would obviously want to say that you only had a 1% dropout rate instead of 100% increased drop rate versus the previous normal rate.

On the other hand, if you are a competing medical school, you would want to exploit that 100% number. You would want to do so because it makes that school look incredibly bad. However, that medical school could also do it to your competing medical school. This is the problem with having two true percentages of the same representation. The ones that represent you the best will be displayed on yours, while your competition will display it as the one that serves you the least. In other words, your average consumer actually has contradicting information that doesn't get the point across. The statistics failed to display the data properly because you have two conflicting point-of-views that are both correct.

## Why You Don't Ever Want to Misrepresent Data

Misrepresenting data is a very dangerous game to play because you're often dealing with people's decisions. You know, with the medical school you're talking about a person who is about to spend eight years of their life and an obscene amount of money. Statistics are a vital part of society and yet they are often wielded by marketing agencies as just another tool. Just another way to convince a consumer to go with their product instead of their competitors. Many of them do not even think of the repercussions they're causing amongst their customers.

What they care about is how much money can be made based off of the advertisement that they make. It's not entirely their fault either because most of it is the fault of the people who employ them. If they want their company to do well, they have to make really good advertisements. It is difficult to make really good advertisements but easy to make sabotage advertisements of competitors. It's very easy to find statistics that make your competitors look bad.

However, this doesn't solve the problem but create a new one. You see, customers want to go with the best product. By having statistics that drag every other company down into the mud done by every other company, everyone gets lowered. In other words, it is very difficult to climb back out of the hole because everyone keeps digging it even further. It's kind of like the AAA industry.

EA was notorious these past two years because they came out with microtransactions inside of their video games. They did this because microtransactions make a lot of money, which is what a company wants to have. A company, along with its investors, wants to make as much money as possible. The problem is that they try to force games out, which is an artistic medium, so they can shove those microtransactions inside of it. For a while, the strategy worked and they made a lot of money until eventually it became an unrecoverable problem. You see, EA has seen a drastic drop in continuous customers and it's affecting their bottom line.

Essentially, while microtransactions do make a lot of money, they also tend to make people mad if the game is not worth playing. It also makes people extremely angry if they bought the game and they can see microtransactions are the point of the game. Like in Shadow of Mordor #2, taking the microtransactions out of the game actually broke it.

By showing how much money, using statistics, can be made with transactions EA fell into a trap. They fell into a trap where you could use too much inside of a video game and begin to lose money. We are talking about a billion-dollar business that controls hundreds of thousands of jobs. They fell for bad statistics that misrepresented the data by exclusionary tactics.

The first part of the statistic told them that they make a lot of money with microtransactions. The problem is that the statistic failed to mention that they make a lot of money from a very small population of gamers. This led them into investing more into microtransactions than into the core product. After a couple of years of games that seem to have publicly flopped, they finally see the error of their ways. In fact, as of writing this book, the newest version of an EA game is said to not have any microtransaction in it at all. This is because money is a finite resource and the majority of gamers are not wealthy. Therefore, while you may have 1 millionaire that's playing the game, you probably have 1 million low-income families playing the game. It's important to attract the 1 million low-income families because a few of those can drop off without hurting your business. If the millionaire that you relied on dropped off, you'd close business.

## Why This Occurs

There are two primary reasons why this occurs. The first reason is intentional, and the second reason is ignorance. The actions of Electronic Arts, otherwise known as EA, was that of ignorance. They looked at the statistical data of how they make money and found out that it was mostly through microtransactions. This created their version of "games as service" instead of games as being just games. Essentially, based on misleading statistics, they thought that they could make more money by making games that people spent more money in than more time on. The goal was to get the money that people pour into games during the initial time that they purchased the game. Every purchase after the first 30 days of a game is really just unexpected income and it usually doesn't amount to much. You see, many people want to test out characters and want their characters to look awesome from the get-go. Therefore, they are willing to buy loot crates and cosmetics out the wazoo to achieve that goal. I think the newest Apex Legends shows an example of this with the character known as Wraith. Wraith has a very specific skin that is highly rare but allows you to use a knife whenever you attack somebody without a gun. According to some sources, it costs $500 to get that character from just the Apex Legend crates. It doesn't give you any extra damage, it just looks cool. In my time of playing the game, I have met several people who have that skin. The average video game itself costs $60. Therefore, if you can get those that purchased the game to also try and get the skins, they want from loot crates, you can make an obscene amount of money.

The only problem is that people begin to understand your tactics. They start mitigating how much money they spend on loot crates because they know that they are encouraging you to do it even more. So, you only have the wealthy that actually participate in The Loot Crate game. You might think that these companies don't mind that. The problem is that when you close everything off with a Loot Crate game, people don't want to play your game. Therefore, the people who might spend $5 or $10 on your next battle pass are not there anymore; you're not going to get that money. That's because they've decided to move to a game that doesn't have microtransactions. The more and more microtransactions come out, the less profitable microtransactions become. Therefore, what is likely going to happen is that Electronic Arts is going to have a microtransaction game that comes out every few years with solid single players in between. This will give them the best option of both worlds. In fact, with EA's announcement, it also seems like it's already headed in that direction.

The next one is intentional, and it's usually done by either the competitor or the person against the competitor. The competitor will take a study that your company did and that got published on one of the many publication websites. They will then look through it for any dirty laundry they can use against you. They will then misrepresent you by claiming that you fudged your numbers, did an inaccurate study, or even misrepresented the data itself. Lastly, they may just employ an article writer to create controversy by tipping them off to the misrepresented version of your data. The person against the competitor will go in the opposite path. They will try to look for anything positive in the study and try to clear up any misunderstandings. They will also go about trying to mitigate any potential misrepresented data points that's inside of the data. However, we have definitely talked about that at length at this point.

## What Happens When You Misrepresent

You might wonder why understanding something like the two true percentages matters. In 1995, a specific type of birth control pill was reported to have a 100% increase in likelihood of blood clots. In other words, the average consumer confused the idea that the pill had a 100% chance of giving you a blood clot. In reality, the issue was that the first generation of that pill caused blood clots 1/7,000 of those who took their pill. The second-generation cause blood clots in 2/7000 of those who took their pill.

Thanks to this very misleading statistic, several trustworthy articles were written around it. In fact, it caused somewhat of a scandal as one might say. As a result, more young women took the first pill over the second pill as a result of trusting these articles. Even worse, the number of young women taking that pill stopped taking it as a result.

The consequence of this is that the amount of unwanted births skyrocketed as a result.

In other words, due to the fact that the data was misrepresented; and it was spread via the news, children were born. A massive number of unwanted children were born into the world simply because data was misrepresented.

## Why Articles like These Come Out

There are a couple of reasons why articles like these exist. Perhaps the first and most common reason is because the person just doesn't like the company they're talking about. You see, there is an old adage that you can't please everyone. This is especially true of the online world. The online world is where the vast majority of statistics are being displayed for the average user to see. Knowing that combination, one can derive that most of the people creating those statistics are online themselves.

The second is simply researchers that are new to the field. Researchers that have been veterans of the field have also seen their statistics being misused. They understand the issue of misrepresenting data. Newer researchers into the field of statistics tend to think it's not that big of an issue until they read a book like this. This is because they have yet to have their statistics be used against them. They are ignorant to the ways of how media feeds off of issues and conflicts, which means they are far more likely to make such a simple mistake.

The last one is usually that they are being funded. News organizations are owned by people who have stakes in other companies. Specifically, the head honcho of the owner of the news organization usually owns a pharmacy or a research outlet. Usually, they are millionaires and sometimes even billionaires. They don't want their own news service going against them. It's unethical to say the least, but it happens more than you know. Thus, when they can use statistics against competitors in other industries, they will use it to their full capacity. They will try to bend and manipulate their news to conflict with what their competitors actually are.

This happens quite frequently and once you understand what the person who owns the news organization also owns, it becomes very evident. Since news is not something that is regulated by the government, they have full leeway to do it and never tell you anything otherwise. The best journalists that are out there often tell you if they are in collusion with other companies; if they have something to lose if that company is hurt. This is to point out that the journalist may be biased. However, that is rarely done nowadays.

## Never Trust Anything That's Not Sourced and Uses Statistics

This is the part where I practically quote every single academic institution that has ever existed. Well, existed within the past hundred years of the United States of America. You see, it's an academic standard that academic papers have sources that are credible and can prove their work. While it is an academic standard for academic institutions, it's not really a standard for people who write articles online. The funny thing is that most of the article writers that are part of the news organization don't source a lot of their information.

Now let me be much more clear about that. They do source their information. What I mean by source is not the idea that they got their information from somewhere else. Instead, what I do mean is that they include a bibliography or a citation reference at the end of their articles. If you look at the average news article, the most common thing that they will do is they will hyperlink just the item that backs up their story. That doesn't mean that they'll hyperlink any other factoids they include inside of their article, just what backs up their story.

This is due to a few factors that many in the journalism business no longer do. If you looked at articles from BuzzFeed or Vice, these articles are often written by new-age journalists. These journalists link to things that they think people might want to know more about. In addition to this, some journalist only ever links to other things so that the Google search engine prefers their article.

This is known as engine search optimization. Google developed a way for it to rank the websites that it links to. This is because of the porn industry and hackers who broke website rankings with a little bit more than useless links. In the early days, Google would associate websites to keywords. The average person wouldn't do anything more than do their best to fit the definition of their keyword. However, that particular industry and hackers figured out you could just stuff keywords into hidden elements of the page. This would cause what's known as keyword stuffing, which is where you go into the HTML of the website and you see like 30,000 lines of just the same repeated keywords. Needless to say, Google kind of needed a different way to measure their websites.

Search engine optimization is its own topic and it is fantastic to use for companies, but difficult to understand. Part of what makes up the value of your section of a website is the authority that's behind it. This is known as backlinking because if you include links to other websites that are high in authority, that authority value passes on to you. This places you higher on the list of links for Google searches for specific topics. This is ultimately what the journalists are trying to do whenever they linked to other items.

Knowing that, you can understand why they don't really want to link to PDF documents or similar items that are just downloadable and don't really have any authority to them. They want to link to credible websites to boost their own website. This leads to a situation where you only get half of the information that you normally get out of an academic paper. Sometimes, it isn't even as high as half, you may get 1/4 or none at all. Therefore, the best way to trust statistics coming from news articles is if they attempt to cite every fact that they bring up that's not a common fact. You don't want to trust organizations who use statistics to justify their points. These are the organizations that will look at perfectly valid statistics and bend them to fit their narrative. BuzzFeed, from my experience, has done this quite a few times. Another example is Fox News. They are a little bit better than they used to be, but in the beginning they used to bend statistics quite regularly in my opinion.

## Use of Incorrect Scaled Graphs to Prove a Point

### Small vs Big Plot Problem

You know what's strange? People are more afraid of a knife than they are of bacteria. The reason why I say that this is strange is because bacteria is responsible for far more death than a knife. We know this because we all get sick and some of us die from being sick. In fact, before we had our field day in medicine, we all used to get sick and die eventually. If it wasn't for war you tended to die of sickness before old age. It was actually a remarkable feat to make it to old age.

The reason I bring this up is because you can make it seem like knives are a bigger problem than bacteria. First of all, you select a well-developed country like the United States of America, Canada, or the United Kingdom. This gives you a population that suffers less from bacteria. Then you frame the statistic so that in the act of an incident with that item, the incident led to death or permanent disfigurement. It is important to keep in mind the fact that we increase the definition beyond death.

82

This is called framing the statistic in a biased way. We already think that you are not likely to die from bacteria and certainly not likely to experience permanent disfigurement. According to the FBI, knives are five times more likely to be at the root of murder than assault rifles. Therefore, you have not only the homicide rate where knives were used but also the assault rate that coincides with permanent disfigurement.

However, this is misleading. Bacteria, on its own, is far more deadly. It has killed tens of thousands in this year alone versus the knives, which are likely to not even break 10,000. Companies do this to devalue several different things. First of all, by being able to redefine the parameters of the statistic, they're able to reflect a bias that favors them. Secondly, the devaluing targets their competitors. This also means that not only companies do it but also individuals seeking a political point or an activist point. For instance, by lumping in rape with sexual harassment, one can display a graph where women's issues are far greater than men's. Furthermore, if you include things like "cat calling" and unwarranted flirting in with sexual harassment, you can further diminish the man's stance. While I won't get into that real-world study that was conducted, it shows you the damage that can be done with such a technique.

The truth is that this is used on an almost constant basis because of how effective it really is. In fact, there's a little psychological trick that some of them throw in and that is to color the small one in blue and the big one in red. Since the days of the first ford vehicle, we have known that red is a special color that grabs our attention. By making the bigger part red, you make it stand out more to the average onlooker.

Let's say that you are running for president and you want to show how bad your competitor is versus you. Let's say that you were anti-war and you've been in Congress for about three years. What you could do is you could go through and pick all the anti-war bills that you signed on to. You could count those up and then compare them to a competitor count. The competitor count would be made up of every bill that led to war. Not necessarily pro-war, but a bill that a had an effect in the positive towards war. What this does is it gives you an inordinate amount of extra data you can shove into your enemies' column while keeping yours small. This would benefit you if your electorate is anti-war.

You could maybe label it "Bills for War" and then promote the statistic to better your position. Likewise, your competitor could do the same to you on maybe a different subject or even the same subject. The point is that many people and companies use this technique to make themselves look better during competition.

Does anyone remember the graph that Intel showed at the I9 viewing? Perhaps you remember the statistic of better-quality games that AMD showed off with the Ryzen series? No? You would not be the only one because those statistics are made purely for that event. Those statistics tend to get buried and forgotten until someone notices that the company definitely lied.

## Many vs Few Plot Problem

A very close sibling to this form of deception is the many vs few plot issue. Sometimes, companies like to use Scatter Plots to show trends. This allows them to correlate an item to two different variables.

Therefore, if you're looking at the age of a smoker versus how much they smoke, you can put them as a plot point. It allows you to collectively gauge your data by looking at a cloud of data points. Unlike a bar graph, a plot point shows trends and trends are really good for advertising why your company might be the way to go.

However, plot graphs provide a very unique opportunity for companies to deceptively use multiple variables to create a narrative. Let's say that a company wants to sell people on the idea of BDSM. Now, the average person cringes at the idea that a company would want to do that but they are out there. The company has to convince the average consumer that it might actually be a good idea to try out the company.

Therefore, what they might do is they might graph age to relaxation. Therefore, those that participate in doing the torture get a relaxation feeling from it and are of a certain age. By creating a graph like this, the company is able to persuade the consumer that they will get a relaxing feeling from taking part in this. They don't label it torturer relaxation, they just label it BDSM relaxation. In other words, they only take the opinions of the torturers and not those that are victim of the torturers. This creates a lopsided view of the picture that shows the company in a good light because they are providing a form of relaxation.

You might think that such an example is ridiculous, but many companies actually do something very similar to this. Let's say that we're going to build a nuclear power plant in a city, something that actually happens every once in a while. The citizens are concerned that the nuclear power plant might cause problems in medical fields and tourist attractions. As a company, you don't really want to hear that the public is against you. Therefore, it's time to perform two surveys at the same time. The first survey is going to see whether people like the view of a nuclear power plant. The second survey is the effects of a nuclear power plant in a city that actually has one.

For the first survey, you are just surveying how people like the look of a nuclear power plant. However, the people that you survey just so happen to be in the field of nuclear research. These people are more inclined to like the image of a nuclear power plant. This is because it represents a scientific achievement for long sustainable power.

The second survey is of a city that benefits from nuclear power and you're going to do a survey within the nearest 20 miles. That should give everyone enough radius to believe that they're safe. What you don't include in that survey is that the nuclear power plant is about 15 miles from the city. Then you place the scale at 30-50 miles. This would create a graph with minimal medical issues while also garnering quite a bit of answers to display on a scatter plot.

Now, if you color the first survey plot points green and the second one blue, you can take advantage of some psychology. Blue is a calming color while green is often associated with something that is good. Having your data in these colors brings a lighter mood to your topic. Additionally, because you manipulated how you collected the data, your data fully backed you up. This is an example of how scatter plot data can be used to manipulate the narrative behind the data. You set out to survey people but only a few of them that are likely to show you good. You set out to survey a large area that had a few people. Thus, you had "many" data points with "few" counter-data points and is a Many Vs. Few Plot problem.

## Definition Problem

You see, the majority of the trickiness around marketing has to deal with definitions. It's not necessarily the data but how the data is both defined and the display of the data is defined. Companies are held to the letter of the law, which means you cannot purposely mislead consumers. However, it's not misleading if you over generalize. It's not misleading if you're technically right. This small loophole in the definition of misleading is where companies try to shove most of their crud into.

As previously mentioned, there are two primary types of defining data that obfuscate what it means. The first is how the data is defined itself, which is broken up into three different categories. Defining data is determined by the way it is collected, where it is collected, and how it is collected. For instance, let's take smoking as an example. You might want to argue that there are no negative effects of smoking, but we all know that's not true.

However, we can manipulate the first parameter of our data and that is the way that data is collected. The way that doctors prove that it is horrible for us at the most basic level is to do an absorption test. Instead, what we can do is we can survey people who've recently visited a doctor that are smokers. This allows us to gain access to people who are relatively healthy that also smoke. We can then define it as smokers that are healthy versus unhealthy according to primary care visits.

Now for the next part, which is controlling where it is collected. For smokers, you don't exactly want the low end of the income scale. These people are very much not likely to visit the doctors on a regular basis and not exercise. Therefore, they would just not be healthy as a result of being low income. This would be where we would want to talk to the middle class but not necessarily the upper middle class. With higher brackets of income, the more suspicion you draw for asking pointed questions. Therefore, the middle class is much more likely to visit doctors and not ask questions about your questions.

The last selection is for how we collect our data and for this one we would want to collect it personally. The reason why we would want to collect it personally is because this allows us to pre-screen our subjects. On viewing, you can usually spot unhealthy individuals from healthy ones. At least, the ones with debilitating diseases.

The second is how you actually display the data because displaying the data can play on the human mind. We've already explored how making things small and stand out less confuses the data. However, there are ever more inventive ways of displaying your data so it meets your needs. Therefore, let's continue with the smoking analogy. Now that we've got our biased data about healthy versus unhealthy smokers, we need to display it.

The easiest way to display the data would be in a bar graph because it's a this vs that. However, we can actually add a narrative depending on how we previously gained our data. Let's say that we obtained the income bracket as sets of thousands as labels so we could categorize it. If we see that the lower-income individuals seem less healthy than the wealthier individuals, we can use that. What we can do is we can show that original straight comparison. We can then show a comparison between high-income and low-income brackets. What this does is it creates a narrative that suggests lower income bracket individuals experience health issues. However, it does a second item and suggest that it is wealth that determines health not smoking. If this is starting to sound familiar from something in the past, it should. It's very similar to what the tobacco industry did before it got slammed by the health department.

However, this was just one way that an example like this could be carried out. Companies and individuals do this all the time, but there is a key note. They usually only do it when there is a motivation behind it and not when it's just straight science. For instance, Gatorade might want to show that their products are good for you and ignore the effects of sugar. The Texas Roadhouse may want to focus on customer appreciation verses the ill effects of saturated fat. Your local bed sheet maker may want to focus on the visuals versus the sustainability of their low-quality cloth. Almost all of the statistics meaning to be technically right are usually in line with some form of motivation. Therefore, when you see statistics surrounding monetarily based or activist based entities, you should be especially wary of how they go about doing this. This is where you will see the most misleading statistics.

## Use of Semi-attached Figures - Correlational Statistics

The last bit that I want to talk about here is the use of a semi-attached figure. It doesn't require much but I believed that it deserved its own section. A semi-attached figure is a figure of a number that's misleading due to some key information that is missing. In other words, it is statistic that's related to the situation but doesn't actually prove anything.

Therefore, let's say that we have a bunch of white nationalists that like the My Little Pony series. We could then show a graph that shows the amount of time that white nationalists spend on series like My Little Pony. We then write an article or an academic paper suggesting that those who watch series like My Little Pony are highly likely to be white nationalists. This is not true, which is why it is known as a semi-attached figure. While it is true that a bunch of white nationalists do like My Little Pony, not all adults are white nationalists and not all the adults that watch My Little Pony are either. They may even just be parents of kids who watch the show. However, I've used the statistic to incorrectly prove that adults who watch My Little Pony are white nationalists (which isn't true).

# Correlation Causation Fallacy

## Understanding the Difference

### Correlation

Correlation is somewhat difficult to define for someone who doesn't understand what it is beforehand. The best way that I can explain it is that you attempt to draw a connection between something that seems to be happening in two variables. For instance, let's say that the price of ice cream increases as the rise of unemployment increases. One could say that this is a causal factor but without concrete proof, it's a correlation. It just so happens that the rise of gas also happens at the same time as the rise of unemployment. Causation can only be proven when there is solid evidence that A happens because of B.

A lot of things in this world correlate with each other because of the butterfly effect. In fact, theoretically, one could say that everything causes everything else. The problem is that doesn't really get us anywhere. If we try to say that a tuna fish on the bottom of the sea is the cause of dull razor blades, it doesn't really solve anything. By distinguishing the difference between correlation and causation, one can begin to find the source of problems or solutions. This is the key crux of why understanding correlation is important.

## Causation

Causal factors within the realm of statistics has to be solidified before claimed. What that means is that there needs to be evidence that supports the claim. Therefore, the way that you find causation is by exploring correlation. Let's talk about that gas example. To posit that the rise of gas prices causes a rise in unemployment, one would first denote the correlation. The only way that you could prove causation is if everyone who was unemployed (or most of them) said it was because they couldn't afford the gas. While it may not exactly be the only way, it is certainly the best way to prove causation. Thus, when you are trying to find causal factors, you always want to look at correlations. However, there are some confusions as to whether something is causal or correlated.

## Dependent vs Independent Circumstances/Variable

This is where understanding dependent and independent circumstances comes in. A dependent circumstance is where something happens based on something else. For instance, we call the causal factor of rising gas prices to rising unemployment a dependent circumstance. It is only because gas prices were rising that unemployment happened. However, if we had just dealt with the correlation of gas prices to rising unemployment, we would say that unemployment is a dependent circumstance. In fact, the rising gas prices is also a dependent circumstance. Unemployment can be caused by a lack of available jobs, a lack of qualified candidates, and a slew of other reasons. However, unemployment is always caused by something else and thus it is a dependent circumstance.

Independent circumstances are much rarer and are usually the causal factors that you find. For instance, an independent circumstance is the limitation of how much gas there is. The limitation of how much gas there is defined by itself. While one could definitely argue that finding more gas would lead it to be a dependent circumstance, most of the time it is an independent circumstance. In other words, if you try to find the causal factor of why there is a limit, it is not something that can be changed. This is what it means to be an independent circumstance. You can't just create gas out of thin air, it took a lot of time and special circumstances to do so.

However, there is a different portion of independent circumstances that also affect its own definition. For instance, let's say that workers were not affected by gas but were affected by the lack of pay. Unless you can find two or three different reasons that predominate why that position pays less, lack of pay is an independent circumstance. The lack of pay is a separate circumstance from a rise of unemployment. It's its own little issue in other words. This is where people get confused by causal and correlated factors. Causal factors can be dependent or independent, but almost all independent factors are causal factors. Let's run through some examples of common correlation vs causation errors.

**Examples of Confusing Causation and Correlation Events**

**Low Quality Tools and Low Yields in Farming: Causation or Correlation?**

*The Story*

Let's say that you performed a study as to why farming in your area wasn't doing so well. You wanted to account for the common items that farmers had to contend with. Therefore, you measured the quality of the seeds, tools, and work hours put in. You want to measure the quality of the seeds to determine if crops are not growing quick enough. You want to measure the quality of the tools because maybe you could harvest more. Lastly, you want to measure the work hours put it in because maybe the farmers are missing out.

At the end of the study, you find that the seeds are of an adequate quality. You also find that the work hours that they put in is about 40 hours a week. Lastly, the only thing that you notice is that the farms that have a poor harvest also seem to have old and rusty tools. At this point, you have to decide whether the older rusty tools are a causal factor or correlated factor of the poor harvest.

*Devil's Advocate*

*Correlation*

It would seem that the correlation would be the best answer to go with because you haven't checked the soil. What if the soil that these farmers are using is inadequate for the seeds, which would cause low harvest. You could argue that this is a correlated event because you don't have direct proof that this is the problem. Since you have already measured a low harvest with low quality tools, you could go ahead and say that this was a correlated event.

*Causation*

On the other hand, it kind of seems obvious that if you have poor tools a poor job will be done. If your other two variables show no signs of being bad, it's got to be the third one right? It's definitely not a coincidence that a low harvest would be produced by having inadequate tools.

*What it is*

In this situation, this is nothing more than correlation. You see, yes it kind of seems obvious that this would be a causal factor, but that's not how statistics work. Most people would see this as a causal factor because in their mind it couldn't be anything else. The problem with causal factors is you need evidence, which means you need to reproduce this. By saying that you saw that low-quality tools just so happened to coincide with low yield farming, you have stated a correlation. In order to prove this correlation is a causal factor, you have to set up a scenario where you replicate the poor tools to see if you get a low-yield harvest. This would require testing the same farms with low and high quality tools.

# Funny Liquid Can and Bacteria Dies: Causation or Correlation?

*The Story*

Let's say that you've left out a jar of jelly that you thought was empty. You were in the middle of making a sandwich, thought the jar was emptied and figured you throw it away later. The only problem is that you accidentally left that jar of jelly sitting there for over a week. You left it there because it's the office eating area and you don't often make your own food there. No one happened to notice the empty jar in the corner.

When you come back and notice you haven't thrown it away, you look in the inside and see a weird fungus-like white haired thing was on the inside. Now, you had found out about the jar, you felt so horrible that you left it out. However, this was the first time you had ever seen something like this in your entire life. Now you had a debilitating question that plagued you're ever questioning mind. What caused this thing to grow on food you left out?

*Correlation*

As of right now, you're probably feeling very prepared to call this a correlation. After all, you don't have concrete evidence that just leaving it out caused this issue. In fact, it could have been the manufacturer that put something weird inside of your jelly jar. It could have been the air around the jelly jar that caused this issue to happen. You essentially don't know ahead of time what could have caused this issue. You are also probably finding it a little bit hard to resist automatically assuming what it truly is because you've been caught off guard before. That is if you got the previous one wrong.

*Causation*

It would seem kind of obvious that if you leave a jar of food out that it will get mold on it. If you had not sustained belief that you had never seen something like this, it probably was very difficult not to say that it was the cause. However, it's kind of limited as to what the cause is because there's not much to affect it. It's not much more than just assuming what it is, primarily because you have seen it in the past, which creates its own sort of evidence.

*What it is*

This one is actually a causal factor, which might either throw you off guard or you're shaking your head like you already know. However, you likely didn't account for every variable. You see, the United States of America has standards by which food is allowed to be produced. One of those standards is that it can only contain so much mold. In other words, the manufacturer could not have affected this process because they are not allowed to thanks to the FDA. In fact, it creates a background of evidence that suggests the manufacturer had no hand in it whatsoever.

In addition to this, you also have the fact that the jar was almost empty. You had been storing the jar the way it was previously supposed to be stored beforehand. This meant that so long as it was in its controlled environment, nothing would happen. What this does is it rules out any effects caused by manufacturing and any new variables.

The controlled environment has the same air as the one that you placed the jar in. When I say rules out, what I really mean to say is that the probability of it being those is extremely low due to standards.

What this also means is that the only thing that was not accounted for was the placement of the jar. The average statistician would then look at the situation and say that the placement of the jar had a high probability of being a causal factor. As mentioned previously, you can only prove causation with evidence. Therefore, one would just need to replicate what they thought caused it to see if it happened again. However, because of how controlled extraneous variables were before the situation, there is enough evidence to support it being a causal factor.

## Chicken Soup is Good for the Sick: Causation or Correlation?

*The Story*

I think at this point everyone has heard, most of everyone, about how good chicken soup is for you when you're sick. Most of us are introduced to this little factoid as a child when we do get sick the most.

Our parents recognize that we're sick and we get a bowl of high sodium noodles in water and chicken flavor.

However, when looking at this from a causal versus correlated event, it becomes a little gray for most people. The problem lies in the fact that most people are told this from birth and so they had just assumed that it is a causal factor. Since there's not much of a story behind the chicken soup, let's explore the two different options.

*Devil's Advocate*

*Correlation*

As we mentioned before, everything starts out as a correlation. Therefore, when you don't have direct evidence that something is a cause, it starts out as a correlation. That is if the success of the chicken soup coincides with lowering the time you are sick.

*Causation*

The average person would assume that this is a causal event primarily because it works. The easiest way that the average person

106

would explain this would be from anecdotal experience. In other words, the average person would say that when they are sick that chicken soup helps them. The problem with this is that it's anecdotal and statistics are not based on anecdotal experiences a good majority of the time. In fact, if statistics were solely based on anecdotal experience it would cause a lot of problems.

However, when you have a very large group of people saying this, it expands beyond anecdotal experience. This is because when a vast majority of people agree on one thing, it becomes your evidence for a causal factor. Therefore, if people with a stab wound say they survived longer because they didn't remove the knife, it becomes supporting evidence that you shouldn't remove a knife after being stabbed. This is the best argument for chicken soup being good for sick people.

*What it is*

This is in fact a dependent causal factor because of the evidence surrounding it. It's not a causal factor just because it happens to a lot of people. The causal factor is dependent on the sodium that's inside of the chicken soup that's made of liquid.

When you are sick, it has been proven that you need a lot of liquid because your body is going to use it. It has also been proven that a lot of sodium can retain quite a bit of liquid. This allows you to use liquid that would have normally escaped the body. In addition to this, warm liquid is known to relax the body from various massages and comfort foods. Therefore, this particular food happens to be a causal factor because it just so happens to have salt and warm liquid. The chicken soup itself is a correlated factor, but the chicken soup being good for the sick is a causal factor dependent on other things. Therefore, if the chicken soup was low in sodium and water, it would not be as effective. In other words, it's only a causal factor because it fits certain parameters.

## Why the Average Person makes these Mistakes?

## Everything Starts as a Correlation

I've probably already mention this, but everything that just so happens to take place at the same time starts as a correlation. Therefore, if you eat a taco and shortly after you have to go to the bathroom, it's a correlation. If you have "Chinese food" and you're hungry 2 hours later, it's a correlation. The problem is that a lot of people assume that correlation is causation.

In fact, the majority of "old wives" tales are correlated events. Therefore, the myth that you will be granted protection from bad luck if you throw salt over your shoulder is correlated. Someone picked up the habit because they noticed it tended to work. We now know that it has nothing to do with that. Well, it's a correlation but we have yet to prove causation. A huge hurdle that many people have to get through is overcoming that confusion of correlation and causation.

Therefore, when someone goes to open a door, they know that the door is opened with a combination of turning the handle and pulling the door. However, it's only when somebody else replicates it multiple times that it is a causation. Until that evidence comes into existence, it's correlation even though it seems obvious.

This is the number one type of correlation vs causation mistake that a person makes when it comes to statistics. The brain is taught from practically day one that when you see something happen more than once, it's causation. In statistics, it's only causation when you can replicate it or there is enough logistical evidence. Logistical evidence could be numerical data, well it's almost always numerical. Numerical data is often what is used in place of replication.

## Evidence Proves Causation

Causation is confusing when it comes to evidence for those who are new to statistics. This is because our initial version of evidence is that something is there to prove it. This is not always the case because there's circumstantial evidence, dependent evidence, and independent evidence.

Circumstantial evidence is something that only happens in a specific circumstance, which means it's a type of dependent evidence. However, circumstantial evidence deserves its own category because of how complex it is. A good example of it is if you sit in your chair and it squeaks, you have created a circumstance. Your specific chair and the way you specifically sit creates circumstantial evidence for the causal factor of the issue. Therefore, your specific chair is a dependent variable and your sitting posture is a dependent variable. Multiple dependent variables create what's known as circumstantial evidence. In other words, it will only ever be true given that certain circumstance.

In addition to that, you could have more than just that one circumstantial evidence. For instance, you are not the only one that sits in a chair and has it squeak. Multiple people around the world sit in chairs that squeak, but almost every single one of them is slightly different in some way to yours. It's known to that person that that specific circumstance is caused a specific way, which is what makes a circumstantial evidence different from dependent evidence. It's evidence that's dependent on circumstance. Therefore, it's true but it only happens given varying parameters of the same dependent variables.

A dependent variable can seem like it should be obvious but it's not always. For instance, as with the example of the empty jelly jar on the counter, it became a dependent variable because of in existence factors. It was dependent on the Food and Drug Administration holding manufacturers to standards. This was a factor that was introduced after the story was told to you. This is often how dependent factors are found. As mentioned before, there can be multiple dependent variables that collectively provide a causal factor.

This is because causation can have more than one. Therefore, let's look at a bird in flight. One person could say that a bird can fly because it's so light. Another person would argue that it's because it's got wings, while the last person argues it's gravity. In fact, it's actually all three in combination that cause the event. The bird is light so that it's not going to sink in the air, it has wings so that it can propel itself upward, and it works against gravity using air resistance. The weight of the bird is dependent on the environment. The wings of the bird is dependent on the evolution of the bird. And the ability to work against gravity is dependent on the previous two. This is what it means to have dependent variables creating a causal factor.

An independent variable/evidence is usually a constant. What I mean by that is that an independent variable is not based on anything else. The greatest example of an independent variable is the loss of gravity in space. It is something that just is and it can be dependent on whether other circumstances are there, but the fact that there is no gravity in space is an independent variable. It's a constant that only changes when something new is introduced. Independent variables are actually somewhat rare because they are solidified.

However, there are circumstances where independent variables are dependent variables. For example, let's say that you are doing a study on autism. Autism itself is dependent on the patient as for its severity and its existence. However, the existence of autism is an independent variable. In other words, it's something that you would have to look into separately for it to be dependent. Therefore, in your study, the autism is independent of its dependencies for your study. This is what's known as treating a dependent variable as an independent variable for the purposes of study. This happens quite a lot.

## The Idea of Common Sense Gets in the Way

A huge part of the problem is that there's this thing called common sense that people just assume exists. Common sense is a horrible, dilapidating thing to have in statistics. A lot of people run on common sense and rely on it, but it's not *scientifically* reliable. The only thing that is *scientifically* reliable is understanding correlation from causation. Therefore, most people would say that if you pull on a door handle and the door opens; that's common sense. However, in the world of Statistics *that* is correlation. People also say that putting toilet paper a certain way is also common sense. However, it's really just a correlation of personal experience. It doesn't really matter which way you put the toilet paper. It's going to have the same effect either way.

You see, common sense is something that blinds you from seeing relevant correlations.

Let's say that we are studying a group of monkeys that have managed to get AIDS. What we want to study is how the monkeys deal with their situation. We want to see if the members that have AIDS are excluded or still included into the group. Common sense would dictate that sickly members of the monkey group would be excluded. The monkeys would want to get away from those that are sick, because that's what we as humans do. We try to stay away from those that are sick because we don't want to get sick. This is the problem with common sense, because that's not what the monkey group does.

Due to the fact that monkeys are a highly social group, members with AIDS in a monkey group are included. In fact, they try to take care of the sick monkey as much as possible. This is because they know that a sick monkey taken care of has a higher chance of living. They don't understand the concept of AIDS or getting sick themselves. They just understand that a member of their group is not doing well. Therefore, the other members of the group will substitute for what that member of the group would have normally done. Common sense would have blocked you from ever exploring the subject in the first place. This is the huge problem that is common sense.

The average person will assume something is true based off of their past experiences. This then blinds them from seeing potential correlations and causations in their statistical data. This also blinds the average person whenever they are reading the statistics of others. Therefore, the common occurrence of video games causing violence is still heavily debated. It's heavily debated by those can't read statistics properly.

If you look at a graph that shows a rise in violence along the rise in sales for violent games, the average person will derive that it is causation. This created what we know as the ESRB. There have been tons and tons of studies disproving this fact. However, because of the common sense that violence begets violence, it's common sense that violent video games begat violence in reality. This is not the case, in fact it's the opposite. Violent video games offer an outlet for those who would normally be violent in reality. It's a form of therapy, yet the debate still rages on because of people who can't read statistics properly.

## The Brain and Rationalization

A huge portion of the problem is the way that the brain is naturally built to rationalize data. You see, the brain doesn't like to be wrong and it does a really good job at lying to itself. You see this in conspiracy videos that talk about the face on Mars, which is really just mountain formations. However, because your brain is trained to notice facial features, you see a face even though there isn't. The brain prefers to be right because being wrong used to be harmful.

Let's say that you were a caveman back when we still scrounged around for food. If you are in the middle of a jungle and you see a shadow, do you assume that it is a predator or nothing? Nowadays, we like to convince ourselves that it's nothing. The problem is that the caveman days required that we assume the shadow was something. The reason being is that the brain understood that if we assume the shadow was nothing, it could be a cougar that killed us in the next moment. This creates a brain that doesn't ever like being wrong and is suspicious of practically everything. It creates a brain with trust issues.

Therefore, what happens is that the average person is inclined to jump to conclusions. This is the second most common reason that people fail to read statistics properly. They want to assume that their first reading was the correct reading. In fact, usually the first reading also compounds itself with bias tendencies. Therefore, if one graph shows a rise in income as another shows a rise in violence, some might look at it and assume a causal factor. This is because they are biased against those who disagree with them, which is how the brain was built to survive. It was built to fight against things that was against it. Therefore, if they have a preconceived idea that goes against the statistical data, they will often read it incorrectly.

As I will talk about later, this is a huge portion of why the wage gap is still such a big deal even though it's been disproven. Several economic experts have come out against it basically saying it is an abysmal representation of statistical data. However, those reasons we'll talk about later. Just know that when you try to look at statistical data, one of the things that you have to do is you can't assume anything about it. You cannot derive what you want out of it, you just have to see it for what it is.

This is often referred to as personal bias and it's something that plagues all industries of science. Sometimes you want something to be true when in fact it's really not. It does a lot of damage until it's caught by someone who manages to call it out. Usually, you want to be the one calling yourself out for having that bias but that's usually not the case.

## Correlated Events Can Also be Dependent Causal

Sometimes the data itself is rather confusing and a lot of factors can seem unrelated until they're put together. Let's take the incidence of high violence, hurricane Irma, and high mortality rate of infections. To the average person, one of those is not like the other. It's kind of understandable that during Hurricane Irma, one might expect there to be a high amount of violence. This could be due to panic, less security to control criminal activity, and many other reasons. It would be its own independent factor.

However, the average person might find it rather odd to see a high mortality rate of infections being correlated with hurricane Irma. The truth is that high infection rates are actually somewhat common during hurricanes. The highest infection rate would be considered a causal factor that is dependent on correlating events. You have hurricane Irma, which is bringing in a lot of water and debris. You also have violent events, which usually result in injuries. Dirty water being exposed to open wounds causes infections. Thus, it now makes sense for a hurricane to be part of the causal factor for high rates of infection. The problem is that it's not immediately evident that this would be the case.

This would also be known as correlated events causing a separate event to occur but is not immediately present. Infections usually take time to develop, this time would have happened about a week or two after hurricane Irma. Therefore, if you are a director at a hospital and you are seeing a lot of infected patients coming through, where's the source of the problem? It's not always the patient's fault for getting infected with an open wound. You can try as much as you might, but infected wounds are usually a result of someone not even knowing that they have an open wound. On the other hand, some people see that they have a wound and let it be fest until it's almost at the point of no return. It's a messy situation, but if you trying to find the cause of that situation you might look at their activities.

Their activities would dictate that they were in the hurricane and they may have been injured in that hurricane. In addition to activities during the hurricane, they may have been injured after the hurricane but from something caused by the hurricane. The act of having a causal event be dependent on correlated events is somewhat common and can cause confusion in the data. It's one of those circumstances where your statistics are not going to seem relevant until you have an "aha" moment or someone else points it out for you. The problem that a statistician has is showing the causal effect through statistics to other people.

As you might have guessed, showing that there is a high infection rate after a hurricane usually demands a why. Why are these people being infected after the hurricane has already hit? Well, infections take a slow amount of time. However, the average person would assume that they have no correlation because the infections didn't happen during the hurricane. Instead, you had a high rise of infections shortly after the hurricane and thus the average person wouldn't see the connection.

**Anecdotal Changes to Evidence with Sample Size**

Anecdotal experience is anecdotal experience until it's not. As you can see, it's something that can get confusing rather quickly. Many statisticians and scientists will dismiss something if it fits in the realm of anecdotal experience. Anecdotal experience is an experience that is personal to oneself. In other words, it's an experience that doesn't necessarily hold any evidence behind it.

A very common reason why the average person doesn't understand something is because they can't relate it to anecdotal experience. They can't take the data and apply it to something they know in life. This creates a problem because statistics are often used for very vital reasons. For them to try to relate to the statistics is part of the statistician's job. The statistician needs to display the data in a way that the average person can relate to it.

However, if you have a big enough population size that has the same anecdotal experience, it can be represented as a causal effect. This is the weakest type of evidence that statisticians use, which is why they don't like to rely on it. If a large enough group of people agree that a certain experience is the standard, it can sometimes just be taken as true. For instance, we all have the anecdotal experience of opening a door. It can be seen as correlational evidence and then corroborated as causational evidence, but that's just more work. It's something that nearly everyone has experienced and the fact that I can use it as an example is in itself an example.

# Politics

## Trumps' misleading report on immigration link to terrorism

Let's talk about Trump for a little bit because you can't talk about politics nowadays without talking about Trump. Trump has been known to blast numbers all over the place and not be correct about most of it. This leads to a slight problem in statistics that we often like to call bias.

You see, due to the fact that Trump has been proven wrong in the past several times, when statistics come out, they have bias attached to them. When Trump specifically comes out and names a figure from a statistic, only a handful of people believe him right off the bat. It's become standard to not exactly trust most of the words coming out of the mouth of Donald Trump. That's not to say that he isn't a good president and it's not to say that he isn't a bad president, it's just that he doesn't really fact check. I mean, when you're busy running the country you kind of rely on the information that's been given to you by the news and your advisors. That's the purpose of both of those, but it seems that the advisors might not have all the facts straight. While we could write several books about Trump's false statements, this book is mostly about misleading statistics; so we're going to look into one particular claim.

As you probably know by now, Trump also has a history with anti-immigration. Trump, when running for president, specifically focused on job-related issues. While there's not much information to back this up, I can provide an anecdotal context to the situation. I knew a co-worker of mine that was born out of Kansas but raised in Illinois. He would often talk about how it was difficult to find jobs around the farming side of Illinois due to the fact that a lot of Spanish-based individuals were already working those jobs. He wasn't exactly racist but given how he was talking about the situation, he might have come off as such like the media likes to portray the average Trump supporter. He would often say that they were illegal immigrants and that they were taking jobs that normal Americans should have. After all, he was an American and if an illegal immigrant took his job, that illegal immigrant took a job he could have had. I can see the logic that he had. I questioned him about the validity of whether they were all illegal immigrants.

He actually pointed out something rather odd that I hadn't thought about before. He said that most of the workers that worked the fields in the areas he had known where individuals that were sending money back. Therefore, you might have one legal immigrant that brought his friends with him. Those illegal immigrants were working at a discounted rate until that they could get either official citizenship or at least a work visa. In other words, they were working a job to pay the bills while they were working to become a citizen. They had gotten here but now they were working to stay here. I asked him why that was so bad.

He said, "It's bad because they are sending money out of the United States to family members who will then pour that money back in. It devalues American currency. It is like a bucket with a hole in it that has the water going through a tube that pours it back into the bucket, but that tube has holes in it that are hard to notice. America doesn't make any money off of their labor beyond the product of their labor, yet they are still able to buy American items from outside America using technically illegal money. Since it is money earned it illegally, it's illegal money. While one or two illegal immigrants isn't a bad thing, thousands upon thousands of illegal immigrants disposition the people who are already there. If you have a town of mostly Americans and all their jobs suddenly dry up due to illegal immigration, you now have a small town of illegal immigrants and poor Americans. They are poor because they have no way to earn money and that means that most people won't be able to fend for themselves. Illegal immigration is harming American citizens slowly and people are encouraging it. The rich don't care because it's not the rich that's losing work to them, not yet anyway. Give it time and they'll care when it's too late."

At that point, it became quite understandable why many Americans care about the illegal immigration issue. They just want to provide for their families and illegal immigrants take the jobs that the majority of Americans could work at. The average working household would want to prevent something from taking their ability to work at their jobs. It also made sense because these jobs were low-skilled jobs, requiring almost no training at all. Therefore, you didn't need to speak English if only one member of your party could translate for you. However, knowing this, this leads to a bias in the average Trump supporter. If you follow the logic, you would see how your average illegal immigrant would be characterized as an enemy of the working American individual. You could see how there would be a bias amongst low-skilled workers versus illegal immigrants. The truth is that illegal immigrants are just looking for a better quality of life; but everything has a cost.

This leads into Trump's very off-handed remarks about immigration and terrorism. The reason why the link between the immigration and terrorism is misleading is due to something that we talked about before. This is what's known as a semi-attached figure. The way that Sarah Sanders presented the information was that there were about 4,000 known terrorists trying to enter our country illegally. Then she points out the fact that the most insecure part of entry into the United States is the southern border. As you can see, they've proposed it in a way that makes it seem like those terrorists are coming over the southern border. However, they have not officially connected the two together.

Illegal immigrants can come in one of two ways, which is on land or in the air. It's actually quite rare for them to come in via the sea because well that's significantly harder to do. While there may be true terrorists that are trying to enter our country illegally, they may not necessarily be coming over the southern border.

Due to the fact that Trump supporters are already biased against illegal immigrants, it also leads to a situation where such a semi-attached figure is also backed up by the bias of those who would normally care about it: hard working Americans impacted first by illegal immigration.

## CNN's claim that Obama deported more illegal immigrants than any previous president

While it has been long disproven, CNN once claimed that Obama was responsible for deporting more illegal immigrants than any previous president prior to him had. According to the statistics, this is technically true, however, it's also not. You see, during the Obama Administration the definition for deportation actually changed. Therefore, items that would normally not be considered a form of deportation were now included into this statistic.

This misleading statistic has a bunch of variables that cause it to be true and misleading. Prior to 1987, most presidents had a rather low deportation rate. In 1987, a new law came into existence that allowed judges to deport illegal immigrants based on deportable crimes. The law incentivized those in power to deport more than they regularly would. This, by itself, led to a rise in deportations.

In addition to this, the way in which deportations were counted also changed. Therefore, in the past, sometimes officers would just send individuals back over the border if they crossed. However, due to new laws it was required of those officers to now detain and legally deport those individuals. This means that many of the occurrences where the individual would just normally been sent back and not have been deported had to go through the legal system now. This led to a drastic rise in deportations because a lot of them were simple punitive items that would have normally had them turned away at the border rather than be deported. In other words, instead of just turning them away, they detained them into the United States and then deported them back out of the United States.

On top of all this, there was a new quota that was established in 2009 known often as the bed quota. This quota required that immigration officers had to hold an average of 34,000 "peoples" inside detention on a consistent basis every day. This further encouraged deportation because instances that would normally be possible would now result in detention before deportation to fill up the quota. This new law allowed for these 34000 detentions to be counted as deportations.

This brings up the misleading statistic that Obama deported more illegal immigrants than any other President.

As we have talked about in extraneous terms, changing the definition in a statistic can lead to drastic results. The claim that Obama deported more illegal immigrants than any previous president is actually due to legal changes in the system that handles deportation. It's not necessarily the fault of Obama himself but the laws that preceded Obama that caused it. This means that a statistic can be made that misleads the public in believing that Obama was anti-immigrant, which is not exactly the truth. The problem is that it's difficult to argue against without understanding a deep knowledge of the American legal system and the laws surrounding deportation.

## What is the Wage Gap?

The wage gap is a somewhat tricky issue to tackle and a very "heated" one at that. The wage gap itself is pretty easy to explain. A study surveyed a lot of men and a lot of women based off of filed taxes. They went through and calculated how much men made and then calculated how much women made. Therefore, you have an average of how much every man makes versus an average of how much every woman makes. If you subtract the two numbers, women make less. It's usually between 5% to 25% depending on who you're talking to and which part of the world you're in.

As you can probably tell, this is rather flawed in its argument. The primary problem that such an argument has is that it doesn't take into account circumstances. For instance, the most common reason why men make more money than women on average is because they tend to take the more dangerous jobs. They are more likely to be found on fishing crews or mining rigs where the work is incredibly dangerous but pays rather well. You are also much more likely to find them in the skilled trades industry such as plumbing or air conditioning. Pretty much, if it's a job in which you get dirty you will often find a male doing that job versus a female. The problem is that those jobs tend to pay a lot more than the jobs that allow you to sit in a cushy desk or in an office environment.

On top of that, you have the lifestyle choices of men and women being completely different. A man is far more likely to overwork themselves, and in some countries it's actually a problem. Females are far more likely to take vacations. Then you have the ultimate lifestyle choice that men simply don't have and that is the ability to have a child. While we do live in a world that's always striving for equality, it is far more likely that a woman will stay home when pregnant and the man will work extra to make up for the loss. This causes a situation that allows for men to theoretically make more money than women. It's an illusion on the statistical scale because they are making up for work not being done by the female.

## The Wage Gap Crutch

The wage gap is a crutch for many an individual in the feminist ideology. The wage gap allows them to believe that women are still oppressed in developed countries. It's a very simple statistic that they can use to prove their points, regardless of how bad that statistic is as a statistic. That's not what this book is about, it's just that the wage gap itself is an example of a statistic that is as bad as a statistic can get with how much of an impact it has had.

The wage gap has caused several companies to reevaluate themselves to justify themselves in the public eye. The wage gap has also led the way for massive reform in some sections of the workforce where women are at. Finally, the wage gap has been used to justify certain actions that normally wouldn't be okay. In other words, it's had a massive impact on several societies while being debunked and proven wrong. Even when you calculate for the wage you got, the rest of it is not a justification that women are being forced to be paid less.

The wage gap is a truly evil specimen born from statistics. You see, the average individual will look at the wage gap and see nothing more than just the way to go. However, the socially active see something else. It's a tool that allows the socially active to claim victimhood because it's an obvious difference that they can point to. Many of them know that it is actually a statistic that really doesn't mean anything. What they've done is that they've used the statistic insidiously to provide themselves with an excuse to further an agenda.

You see, the wage gap says that even though the society you're in is supposed to be equal, monetarily the sides are not equal. It doesn't care whether the choices that they make cause this to happen in the first place. All they care about is the fact that the gap exists, which means that women are being paid less. This means that they can push to have women be paid more because, obviously, they are being paid unfairly. They don't care if they are being paid less because they choose jobs that pay less, they just care about whether women are doing better than men.

You might think that this statistic wouldn't exist if we were really paying equally. The problem is that the majority of highly dangerous jobs are taken by men and the safer jobs are usually taken by women. What this does is it devalues the dangerous nature of jobs taken by men and endows value into jobs that are safer simply because they are women-oriented jobs. Therefore, they are able to reduce the pay of men and increase the pay of women so that women have more power.

That is the true root of why the wage gap persists in government and socially active circles.

The claim is that women are being paid less for the same amount of work, which is not true. Technically, it's not true but it is true. You see, they ride the fine line of definition semantics. They are talking about the calculated amount of work that went into what women were doing. They're not talking about the same jobs. They're not saying that a female developer is paid less than a male developer. What they are saying is that a female developer makes less than a male crab-fisher that work the same number of hours. This is where it's truly insidious because they are using definition semantics to lie to your face.

Therefore, if you have a male neurosurgeon that works 30 hours a week and a female nurse that works 30 hours a week, you don't have a comparison. According to their logic, they should be paid the same amount of money because their hours match. It doesn't matter whether the neurosurgeon has gone to school for longer and does harder work. What matters is the fact that the female and male worked the same number of hours. This is the wage gap and when you add in other definitions, you can truly see how insidious it is.

**Application Versus Use**

**Good Intentions: Alimony and Child Support**

Child support was originally meant to compensate mothers who had men walk out on them. When child support and alimony came about, they existed as crutches women could rely on when their husbands left. The reason behind it is because being out of a marriage was considered a bad thing. Men would often avoid women who became ex-wives and the same was not exactly true of men. In fact, it was kind of sexist because women were seen as being the bad part of the relationship if the marriage was dissolved.

In addition to this, there was a time when owning land as a woman was quite difficult because women were generally not taken seriously and not given many roles beyond housework and being mothers. The statistics showed broad direct and indirect discrimination of women across society, which brought about a need for these laws.

Having said that, the government tried to give assistance to this particular type of woman. Now, those same statistics of women needing help and the government stepping in has done an incredible amount of damage. Yes, it did help women sustain themselves when they got out of a bad marriage. However, what it also did was it incentivized women to leave their husbands because they could use child support and alimony against their husbands.

Therefore, you would have situations where a famous man falls in love with a woman and the woman then convinced the man to marry her only for her to shortly thereafter divorce the man. Due to the laws of child support and alimony, a woman would be able to claim obscene amounts of money and usually never have to work again. These types of women gave women a bad name, the name of "gold digger". The name "gold digger" specifically denotes a person who goes into a relationship for the sole purpose of divorcing them and getting as much money out of the relationship as they can. In other words, it created a special class of women predators. It is also one of the factors that led to the increase in divorce rate in Western countries.

Now, in an incredible feat of irony, the same thing that was specifically meant to only benefit women is now benefiting men. Men have not been able to keep pace with the increase of women into the workplace because when you add double the amount of labor force into the market, someone's got to lose. Therefore, women who used to be able to divorce their husbands and almost always expected alimony if not also child support are now falling victim to the same set of laws.

The reason why I called it ironic is because now that the laws have been turned on the women, there is a class of women trying to either lower how much males get out of it or rid society of those same laws. Therefore, you now have women specifically marrying into a higher financial relationship for the sole purpose of getting as much money as possible out of it, some men avoiding all women because of this reason, men also becoming alimony predators, and children suffering across the nation because of it. This was all because of the statistics and the government applying those statistics to the law.

## Using Unjust Statistics Against Itself to Achieve the Best Balance

As interesting as I find the reverse of alimony in today's courts, I don't really like alimony itself. Alimony represents an attempt to balance an issue from the past. Alimony was developed to balance the issue that women had a hard time finding work when they separated from their husbands. In other words, it was literally built for women during a time they didn't really work. The law still exists and is around. The problem is that so many people benefit from alimony that they don't exactly want to get rid of it anymore. In fact, thanks to the use of bad statistics, feminists continue to justify alimony.

The problem is that when you create a law garnered towards one gender but don't specify that because that the law exists because of sexism towards women, you create an imbalance. Monetary systems loathe imbalance and tend to self-correct or self-explode. Due to the fact that literature on alimony was made to be not sexist, men have recently started to gain alimony. It actually represents a key moment in a small battle between the sexes. As previously mentioned, alimony was primarily made for women. However, because women are more involved in the workplace, alimony has started to affect women badly. Women who have higher earning jobs, but stay-at-home husbands face the repercussions of alimony. In fact, they face what men have faced for years. However, when you create laws you need to realize those laws apply to you. In the 1990s, you could have likely heard of a woman getting alimony from their seemingly wealthy husband. Some of them would be justified by a horrible relationship and a cruel husband. In the early 2000s, the tide of alimony collectors has reversed. While when the law was created it was okay, as time has gone on a lot of females have abused the law. Now that females are more active in the workspace, they are suffering from the husbands that abuse it. It's actually creating a balance within the law itself naturally, but it wasn't supposed to do that.

A lot of laws like alimony are made in good faith to protect individuals and provide a safety net for society. However, as we see again and again, the use of statistical data to make laws; without taking into account the underlying causation; leads to abuse of the laws. In extreme cases, it can lead to a lot of suffering for decades.

It represents a moment where statistics that were justified in creating a law that was imbalanced is used against itself. There was actually another example of unjust statistics being used against itself to achieve the best balance; the BBC. The BBC had decided to do an income equality check to see how many women they were paying less than men. It was all over the news that they would be ensuring equality and making women richer by getting them what they deserved. They did the report and found out they were paying men less than women. It's one of those moments where you know they weren't expecting it. They were expecting that women would be getting raises all over the place, but men ended up getting it instead. In fact, you had numerous feminist outlets calling it misogyny that the males got raises in the first place. They didn't see why the men needed raises when it was the women that needed help, only they didn't need help. Essentially, statistics used to garner an unjustified call wound up being used against itself to achieve balance.

This actually represents the greatest part of statistics because statistics are what the wielder uses the statistics for. You have bad statistics and good statistics, but sometimes bad statistics lead to solving problems that weren't known. Bad statistics often have a negative effect because they don't really solve problems but make them worse. However, in some small cases you have bad statistics resulting in a better balance. Alimony was a one-sided law until it became two-sided thanks to balance. The wage gap that persisted to convince the BBC that they needed to ensure equal pay checks for women wound up paying men more. It is sometimes useful to have a bad statistic because when that bad statistic is put into practice, it amplifies the real problem.

Bad statistics have squashed men's issues into a gutter. However, as more people fit into that gutter, eventually you have a crowd that's much stronger. A crowd of people that tackle an issue with more veracity then if statistics were used to justify it from the beginning. Instead of a small house where men can stay, you have a fleet of houses just pop up. Instead of one lone warrior fighting for men's rights, you have entire law firms specifically dedicated to it. This is because the more you try to quash something without killing it, the stronger it becomes. There is a huge need to help men right now and the feminists are doing everything in their power to either obfuscate it or ignore it. They obfuscate it by saying that men's rights are included in feminism. They ignore it by saying that men don't get violently abused as much as women. The problem is that men can deal with violence but it's their emotional side that they're more vulnerable in. The bad statistic led to the rise of helping women.

However, the bad statistics have led into a population of men helping themselves when no one else would have helped them. In other words, the collection of bad statistics have generated what's known as an overcorrection. Instead of a small following that seek to help men, you have a powerful surge that just comes up out of nowhere. This is an example of the unjust statistics causing a much stronger balance to be made. You have entire law firms for men's issues. You have workshop organizations to help men get through things emotionally. You have hotlines specifically for men. You have groups getting men together so they can socialize. In other words, a huge surge of help has resulted from the long lack of help caused by bad statistics.

# Conclusion

## Statistics that Equalize Importance

Statistics can bring about revelations for people who are not really used to a concept. For instance, when one thinks of prison, they often don't think of rape in terms of the prisoners being victims. The idealized version of prisons is that the criminals that go there are some of the toughest and hardest people there are.

For a very long time, many people just thought of prisons as a "put it away and lock it up". No one really paid attention to the structures that made up prisons. No really asked the question about why criminals tend to go back. In fact, many people just didn't care until they got into prison themselves. This is why one of the most important studies in history is referred to as the Stanford Prison study. In this study, they created a fake prison. This fake prison was populated by volunteers that were paid to be there. They were actually paid quite a decent amount of money to be there. Some of the volunteers were selected as guards of this fake prison while the rest were prisoners. The study specifically reduced the number of guards there were compared to prisoners.

The first day of the prison kind of went expected. The guards themselves didn't really take it seriously and neither did the prisoners. In fact, it was somewhat of a joke. However, by the second day, the situation had changed very drastically. The guards were more authoritative, and the prisoners had begun to show resistance to that authority. As one might theorize, the prisoners didn't really fight back until the situation got unpleasant. The way that the guards solves such a problem was to dehumanize the prisoners. Therefore, the prisoners acquired numbers instead of names. While you could definitely go look at this study, it showed a very important thing.

The study was stopped about halfway through the time frame it was planned. This was because it was such an effective environment that it got out of control very quickly. Even the psychologist that was running the experiment fell victim to it. It very quickly showed the damaging nature that occurs between guards and prisoners. Specifically, it showed that prisoners tended to end up back in prison because it changed their minds to conform to prison. Essentially, it raised an important problem with the prison system and fixes have been put in place for a very long time.

People generally didn't care at first because there is nothing for them to understand. Once the study was completed, people began to care and it brought importance to the issue. Statistics have the ability to bring previously unimportant topics to the front.

It also justifies the actions of companies when it needs to make difficult decisions. For instance, statistics are used on a daily, weekly, monthly, and annual basis to judge employees. Employees that think that they can slack off and not do as much work fall victim to these. In fact, it's a cost versus benefit analysis that most companies use as a form of determining employee value. Prior to such an analysis, people would just wind up getting on everyone's nerves. Essentially, the employee that was naturally seen as useless was eventually fired.

Sometimes, you had a situation where the employee was vital, but people saw them as useless. Once that employee was out of the picture, it brought to attention how important the employee was. Now, statistics have the ability to take what the average person thinks is important and normalizes it to reality. It is capable of revealing hidden problems and dismissing unimportant ones.

Previously, issues either had to be upfront and direct or a personal stake. You had to have had family members go through problems before study was done. You might have taken an interest into something odd and study was done on that. In other words, without statistics, you were really blindly searching what you needed to solve. This person had cancer, but you just blindly tried to solve cancer. We now know that there are multiple versions of cancer and multiple rates of success for cures. This is all done with statistics.

Washing your hands was actually a foreign concept once-upon-a-time. The idea of a doctor needing to wash their hands before treating a patient was not exactly common knowledge. It wasn't until they discovered that the success of treatments went up with washing their hands that study into it began. This meant that statistics helped create the common standard of washing your hands before treating patients. Needless to say, such a measure has gone great lengths into saving many people's lives. Yet, once upon a time, it was not a standard practice and statistics made it a standard.

## Planning People and Seeing the Future

There's a lot more to statistics. Statistics are a good way to see trends, which is another way of saying habits. People have habits and we're always looking for ways to make our lives more comfortable or improved. We have smart weight machines that can analyze our fat content from are muscle content. We have smartphones that can take our heartbeat. We even have electronic cigarettes that measure how many times we puff that peachy flavor. All of these generate statistics and can be used to gather data.

I think a good example of this would be an application that is capable of learning when you set the temperature to something different in the house. Let's look at Nest, which is a thermometer that has artificial intelligence. For the first couple of times that you use this thermometer, it records your average changing of temperatures. Therefore, it will record when you're outside of the house and what temperature your house is. It will also record when you're inside the house and what your temperature is then. By creating a statistical profile of your average temperatures depending on case scenarios, they can do something special. The thermometer can now take over and you don't ever have to worry about it. If it detects that you are on your way home, it will change the thermometer at the right time to save you the most amount of money. If it detects that it's close to be a certain time that you leave, it will then change the temperature on its own.

This simple device just learns how you change your temperature in your house on an average basis. It does this for you and also optimizes it. The purpose of optimizing it is for maximum comfort with maximum savings. Every minute that you can save on electricity, you can do it with this thermometer. The thermometer will grab the temperature from outside and determine if it's appropriate to change the temperature. For instance, it is much easier to change the temperature when it matches the temperature outside. Therefore, instead of saying that your house 77 degrees Fahrenheit at 6 in the morning, it can do something different. It can see that outside its 60-something degrees. You have set it to 77 and you have left the house, but it doesn't make sense to heat up the house when it's colder outside. Therefore, what it will do is it will raise the temperature as the temperature outside raises. This allows you to keep your house colder for longer without using extra electricity to heat up the house.

Now imagine this for an exercise app that has access to your heartbeat, weight, muscle mass, fat mass, water mass, and even your habits. Imagine how effectively it can use your time to get the best exercise without interfering with your life. Statistics is how artificial intelligence runs and it can help improve your life vastly.

In addition to this, you have statistics that are used in criminal investigations. There was one statistician that was responsible for catching a serial murderer. He did this with statistics that allowed him to create a probability matrix of who that killer was based on the locations of his victims. You have statistics that are capable of determining whether devices were used in any illegal actions by the user also.

So, let's say that someone has decided to steal your internet connection. They've also done something called spoofing that confuses your device's address as their own. This allows them to pretend to be you. On that internet connection, they download a whole bunch of illegal movies and files. According to obvious logic, you would be at fault for all of that illegal activity. However, criminal investigators know that this is a practice and so they use statistics to figure out if it's you who's doing it. What they do is they check the activity in which the illegal actions take place versus your daily habits. Therefore, let's say that you go to sleep at 8 at night. The illegal activity suddenly starts up at 9 at night. The agent that's been staking out your home checks to see if you are on the computer but you're not. Overtime, your habits and the illegal activity are given their own statistical charts. This allows them to create a profile of who you are and who the criminal is. Basically, by keeping an eye on you and an eye on the criminal activity, they can prove you're innocent. In addition to this, based on the statistics of what is being downloaded, they can get a scope of the criminal. For instance, if they are just using the connection to download illegal movies and songs, it might be seen as a redistributor or a small-time criminal. If there is a sudden uptick in illegal movies being sold or they somehow pop up online shortly after it's been stolen, they're looking at a redistributor. A redistributor is a criminal that steals electronic items and places it in a location that allows more criminals to receive the goods. So not only does it prove your innocence, but it also helps them catch the real criminal.

Not all statistics are bad, it's just that statistics can be used for good and bad people. It's up to you to analyze the statistics being used and determine whether the person who's using them is doing it for self-gain or for science. Statistics are only as powerful as the people who believe them.

# The end... almost!

Reviews are not easy to come by.

As an independent author with a tiny marketing budget, I rely on readers, like you, to leave a short review on Amazon.

Even if it's just a sentence or two!

So if you enjoyed the book, please head to the product page, and leave a review as shown below.

I am very appreciative for your review as it truly makes a difference.

Thank you from the bottom of my heart for purchasing this book and reading it to the end.